# TAMING THE FLOOD

# TAMING THE FLOOD

A History and Natural History of
Rivers and Wetlands

JEREMY PURSEGLOVE

Oxford   New York

OXFORD UNIVERSITY PRESS

IN ASSOCIATION WITH CHANNEL FOUR TELEVISION COMPANY

1988

Oxford University Press, Walton Street, Oxford OX2 6DP

Oxford New York Toronto
Delhi Bombay Calcutta Madras Karachi
Petaling Jaya Singapore Hong Kong Tokyo
Nairobi Dar es Salaam Cape Town
Melbourne Auckland
and associated companies in
Berlin Ibadan

*Oxford is a trade mark of Oxford University Press*

British Library Cataloguing in Publication Data
Purseglove, Jeremy
Taming the flood: rivers and wetlands
in Britain
1. Wetland conservation—Great Britain
I. Title
333.91'8   QH77.G7
ISBN 0-19-215891-0

Library of Congress Cataloging in Publication Data
Purseglove, Jeremy.
Taming the flood.
Bibliography: p.   Includes index.
1. Flood control—Environmental aspects—Great
Britain. 2. Drainage—Environmental aspects—Great
Britain. 3. Stream conservation—Great Britain.
4. Riparian ecology—Great Britain. I. Title.
QH545.F55P87 1988   627'.4'0941   87–22033
ISBN 0-19-215891-0

Printed in Great Britain by
Butler & Tanner Ltd.
Frome and London

# FOREWORD BY TONY SOPER

IN common with most people, I have always been fascinated by water. Any pond, any stream, and any river draws the instant response of a desire to paddle and peer into the mysterious depths. But that water has to be harmoniously connected with its setting. The pond needs to be fringed with vegetation, the stream used by wagtails, if not otters, and the river must meet the sea by way of water meadows and salt-marsh. Too much of our water courses through concrete, the quickest and most soul-destroying way of getting from A to B. Yet the underlying message of this book—written by a man who knows better than anyone how to tell it—is that the tide has turned and a powerful current is moving in the direction of a more healthy waterscape.

When I first met Jeremy Purseglove he was involved in a crusade to change the way in which mile upon mile of river was straightened and cemented in the name of drainage in order to reduce flooding. Since then he and others like him have achieved a quiet revolution in the water industry. The machinery which destroyed so much riverside habitat altered course to extend and encourage natural vegetation as a legitimate part of the land-drainage engineer's brief.

The story of that breakthrough, which provoked considerable interest in the media, was the original impetus behind this book. But *Taming the Flood* amounts to a lot more than that. Written with passion and scholarship, it is not only about landscapes—ranging from the secret brooks of Shakespeare country to the wind-swept Norfolk marshes—but also combines ecology, history, literature, and politics in order to explain why our rivers and wetlands matter and the ways in which they are under threat.

The future of our countryside lies in your hands. If you care about it please read this book.

*To all those engineers, digger-drivers, and farmers*
*who are taking nature conservation seriously*

✳

HOTSPUR. See how this river comes me cranking in,
　　And cuts me from the best of all my land
　　A huge half-moon, a monstrous cantle out.
　　I'll have the current in this place damm'd up,
　　And here the smug and silver Trent shall run
　　In a new channel, fair and evenly;
　　It shall not wind with such a deep indent
　　To rob me of so rich a bottom here.
GLENDOWER. Not wind! It shall, it must; you see it doth.
　　　　　　　SHAKESPEARE, I *Henry IV*, III. i.

# PREFACE

ACCIDENT has placed me in the middle of a long-standing environmental debate concerning river and wetland management. For ten years I have been employed with a remit to defend the natural world by a water authority, one of whose prime duties, to alleviate flooding, has led it, inevitably at times, into conflict with conservationists. Being pig in the middle, I have been well placed to see the faults and virtues of both sides in the ensuing debate, but I do not claim to have found all the answers to the profound questions which that debate has raised, questions concerning limits to economic growth, the need for a balance between restrictions on those who earn their living in the countryside and safeguards urgently required to ensure that in the next century we will still have any countryside worthy of the name, and the vexed question of how we are going to farm the landscape in future in a way which is both agreeable and economically viable.

All I have tried to do in this book is to assemble a jigsaw puzzle. When the reader sees the completed picture, he may more easily judge for himself the rights and wrongs in the conflict of values concerning river management. In chapters 2 and 3 I survey the long history of river management, during which land drainage has continually been the subject of controversy. In the following chapters, I go on to question whether wholesale drainage is always sensible in terms of practical agricultural economics and efficiency. I then examine *how* land drainage is done, and the way in which an activity which has been, and often still is, environmentally destructive can now be carried out to enhance the landscape quality of rivers, while still controlling their dangerous potential for flooding. Finally, in the last chapter, I discuss the case of wetlands, where it is now increasingly accepted, in the interests of the environment, that major drainage should never be carried out. This last chapter is, in a sense, a history too—a history of the 1970s and 1980s, during which attitudes to wetland management have undergone a major revolution.

Of course, history never stands still, and events will have dated some of what I have written before the ink is dry. None the less, reforms affecting the administration of land drainage and, consequently, the future of our rivers and wetlands will be considered by Parliament only after this book comes out. The degree to which these reforms benefit the environment will depend in part upon the level of public interest and pressure which the issue arouses. If the public actively demands high environmental standards for river management, it is more likely to get them. This pressure can be exerted either through conservation groups or by directly approaching those who manage the rivers in one's vicinity. If the crown jewels of the landscape are stolen, then, when the final history books are written, it will be not just the thief who will be remembered and blamed, but also the slumbering guard.

At the outset, in order to avoid disappointment, I should perhaps list those aspects of the river and wetland environment which I do *not* cover in this book. I do not consider the drainage of upland bogs and hilltops, which are frequently part of forestry operations. Nor do I touch on current controversies concerning navigation on such rivers as the Yorkshire Derwent, the upper Severn, and the upper Thames. Though sometimes linked with river engineering, these conflicts do not turn on the issue of land drainage, and would probably require a whole book to do them justice.

I am concerned in this book primarily with England and Wales, although I also refer to France and Ireland, where major lowland land drainage is having a drastic effect on river and wetland ecosystems similar to our own. I have not covered Scotland, where little arterial drainage is carried out, and where there are no water authorities or internal drainage boards.

In a book which is deeply concerned with a sense of place, I have avoided using the word 'British'. There is no such thing as the 'Britishness' of Britain, whereas a great deal is meant by the 'Englishness' of England and the 'Welshness' of Wales. I hope Welshmen will forgive me, where, for the sake of brevity, I have not always added 'Wales' and 'Welsh' to 'England' and 'English'.

# ACKNOWLEDGEMENTS

I AM grateful to the following organizations who have been unstinting in their help: the Association of Drainage Authorities (ADA), the Council for the Protection of Rural England (CPRE), the county trusts for nature conservation, the Friends of the Earth, the National Trust, the Nature Conservancy Council (NCC), the Ministry of Agriculture, Fisheries and Food (MAFF), the Royal Society for the Protection of Birds (RSPB), and the water authorities.

My thanks are also due to the many individuals who generously gave so much of their time and experience, especially:

| | | | |
|---|---|---|---|
| John Bellak | Robert Evans | Chris Mollart | Francis Pryor |
| Tim Bennett | Brian Eversham | Joe Morris | Trevor Rowley |
| Walter Binder | Nigel Holmes | Nick Myhill | Tim Rudge |
| Richard Bown | Rob Jarman | Neville Neale | Carla Stanley |
| David Brewster | Richard Jefferson | Chris Newbold | Ken Stott |
| David Brown | Brian Johnson | Malcolm Newson | Chris Stratton |
| Dinah Browne | Noel King | Harold and Joan | Richard Vivash |
| Frances Burrows | Andrew Lees | Ody | Max Wade |
| Charles Cator | Jill Lewis | Tim O'Riordan | Stephen Warburton |
| Richard Cooke | Norman Lewis | Keith Payne | Terry Wells |
| David and Judy | Richard Lindsay | Edmund Penning- | Philip Wilkin |
| Drew | Chris Mathias | Rowsell | Gwyn Williams |
| Penny Evans | Loic Matringe | Richard Porter | Dave Woodfall |
| Phil Evans | George McElroy | Greg Pritchard | |

Particular thanks are also due to Chris Garrett for the maps, Rodney Ingram for the illustrations, Anne Baird for the photographs, Anita Shovelton for typing the manuscript so professionally, Geoff Pawling for reading the entire manuscript, and to my wife for her invaluable comments on the text and support at all times. Finally, this book would never have been written without Tony Soper. It was he who first suggested it, and it was his practical encouragement which persuaded me to start setting the alarm clock. All errors and omissions in the text are my responsibility, of course, and the views expressed are my own.

# CONTENTS

# LIST OF COLOUR PLATES

## NOTE ON ILLUSTRATIONS

Except where stated otherwise all photographs are by Anne Baird.
Drawings are by Rodney Ingram and maps by Chris Garrett.

# 1

# *River versus Drain*

I NEVER thought for one moment that my life would become bound up with rivers. It all began when a woman threatened to tie herself to a willow tree. I had just started working as a landscape architect for a water authority, and my duties were to involve planting trees around reservoirs and new office buildings. I had never heard of river engineers until one morning a very harassed engineer rang to ask me to persuade the lady in question not to tie herself to the ancient pollard which he intended to fell. Indeed, he asked me, as the authority's environmentalist, to suggest ways of calming the local community's very vocal outrage at what he was intending to do to their river. The engineer explained his problem. All he was attempting to do was to prevent the river from flooding these ungrateful people's houses. For this, however, many trees would have to be removed, and the river would have to be deepened and straightened. It would cease to be recognizable as the river which the local people enjoyed, but it would become a very efficient drain, so that their sitting-rooms would no longer be ruined periodically, and farmers with land adjacent to the river would be able to grow more and better crops to feed the very people who were complaining. So the case was put: river versus drain.

This book is about that conflict of values, the efforts made in the last few years to come to terms with that conflict, and its implication for one of the major issues of our time—the reshaping of our farm policy, which, with the single aim of increased food production, has transformed the countryside over the last forty years in what is perhaps the greatest agricultural revolution since the settling of England began.

Seldom has the conflict over what a river means to different people been more dramatically highlighted than in the case of the river Stour at Flatford Mill. Here John Constable painted *The Haywain*. Reproduced in calendars, on birthday cards, and in coffee-table books, this painting has become almost a ritualized symbol of the reverence which English people have for their countryside. I remember once dealing with a particularly brutal river scheme, and seeing hanging in the engineers' porta-cabin, which overlooked the now canalized river, a calendar showing *The Haywain*. The connection was not made. In 1984 the Anglian Water Authority applied for permission to carry out a land-drainage scheme on the Stour between Stratford St Mary and Flatford Mill, with the purpose of converting the riverside pasture to oil-seed rape. Mr John Constable, the painter's great great grandson, protested in *The Times*: 'Believe me, rape is what we're talking about.' The water authority responded with its case: 'We are up against a lot of pressure from landowners to do something about the flooding.'[1]

A river is a symbol of changeless change. Overnight, in a flash-flood, it will dramatically move its banks, depositing shoals and cutting new channels. In recent decades, the Severn has been steadily undercutting the riverside churchyard at Newnham. The skeletons of the village forefathers are regularly exposed and then claimed by the river, and now the church itself is threatened. The local diocese has appealed to the water authority to do something about it; but something as elemental as the ever-moving Severn is beyond the resources of even the richest and most powerful water authority. At Crowland in Lincolnshire stands a medieval bridge, stranded high and dry in the middle of the town. The three streams which once ran beneath it have long since vanished, but, at the back of the town, the water still finds its way to the sea, as it has from the beginning of time. William Wordsworth described the river Duddon as something which would always be recognized by succeeding generations:

> I see what was, and is, and will abide
> Still glides the stream and shall for ever glide.[2]

Nowadays, however, we are capable of transforming rivers so that they become quite unrecognizable. If, while waiting in the rush-hour queue at Sloane Square tube station, you chance to look up, you will see a large iron pipe. It is, or was, the river Westbourne. This is the ultimate in man's domination of a river, although many city rivers might just as well be piped. The Rea in Birmingham and the Medlock in Manchester hurry down through their strait-jackets of steel and concrete, unnoticed by the passing crowds.

### Wildlife of Rivers

The endurance of rivers, which is part of what makes them such a potent symbol in our culture, is also precisely the reason why they matter so much to ecologists and scientists. In this country there is probably no river or wetland which is 'natural' in the sense that it has never been interfered with by man; but river systems have two major characteristics which have enabled their wildlife in all its original complexity to survive interference better than most other systems. First, their continuous, linear nature provides plants and animals with an opportunity to move up and down them. In the modern landscape, woods, ponds, and heaths, for example, are increasingly isolated within enormous fields of pasture or arable land; and the other major corridor for wildlife, the hedge system, has, of course, been cheaper and easier for farmers to remove than the river itself. Second, because a river's nature is one of changeless change, forever on the move, the creatures which live in it have evolved strategies for surviving sudden floods

The ultimate taming of a river. The River Westbourne flows in an iron pipe above the platform at Sloane Square underground station.

and disruptions and alterations of the river's course. Broken pieces of many water plants have the ability to root again; others have seeds which float or seeds which resist digestion in the stomachs of birds, and so can be transported upstream. River insects develop wings in the last stage of their life cycle, and dragonflies are known to be able to fly many miles. Indeed, some of our dragonflies regularly migrate across the North Sea. In June 1900 the air over Antwerp 'appeared black' with swarms of four-spotted chaser dragonflies, as they headed towards England.[3] Fish instinctively fight their way upstream against the current, and many water birds and animals have the ability to travel long distances.

Other species are less mobile, and are there simply because a river has always been there. In 1983 a hairy snail was discovered in the Thames marshes near Kew, where its ancestors had lived for the last 10,000 years. It is believed to be the last living relic of the days when these islands were joined to Europe and the

Thames was a tributary of the Rhine, where the same species of snail is still found. Now, within two years of its discovery, the snail's survival is threatened by a Thames Water Authority scheme.

Of rather more popular appeal than hairy snails, perhaps, the dragonfly best represents the ancient life of the river bank, that point where land and water meet, where life began, and where waterside plants still provide a slipway up which dragonfly larvae climb to emerge in their full splendour every spring. In the liassic rocks of Worcestershire and Gloucestershire fossil dragonflies have been found which are not very different from those still hawking along the streams which cut their way through those same rocks on their way down to the Severn estuary and the sea.

Over the millennia, creatures which live in the specialized conditions of rivers have evolved by adapting to these conditions. A babbling upland brook is physically very different from a lazy lowland river, and there are subtle gradations all the way between. These differences are further modified by the local geology, which affects the water chemistry, the local climate, and the particular conditions created by the dominant local plants. Thus a river's wildlife is adapted to, and expresses, its particular local character and that of its different reaches with an almost infinite variety.

Dragonflies are a good example of this. There are upland dragonflies and lowland dragonflies. The Norfolk Hawker (*Aeshna isosceles*) is confined to the Norfolk Broads, while the Brilliant Emerald (*Somatochlora metallica*) is a speciality of Surrey, Sussex, and Hampshire. The 'Beautiful Damsel' (*Agrion virgo*) favours clear, gravel-bottomed streams, while the 'Banded Damsel' (*Agrion splendens*), distinguished from the former by the blue-black bar on the male's wing, is abundant on muddy-bottomed, less acid waters. Where they co-exist, the former tends to prefer the gravel, while the latter chooses the clay reaches. The damsels are among the great sights of the midsummer river bank, along which they flutter, enamelled with peacock blue and green; and it is entertaining to 'read' the physical conditions of a particular stream from the band on a damsel's wing.

A single rock in a stream provides at least four habitats. Algae grow on surfaces which are always wet; the dry top supports lichens; mosses thrive on the wetted margins between the two; and many creatures hide in crevices under the rock. On many upland streams a large boulder will often be the chosen perching spot of the dipper.

Further down a river, common reed is a feature of many watersides. Thickets of fawn papery stems, tender green as they unfurl in the spring, have a specialized ecology all of their own. Even quite small stands may support a pair of reed-buntings, while a larger reed-bed provides a home for the reed-warbler. The latter is often a favourite host for the uninvited cuckoo. Look closer into the reeds, and you will find a world within a world. The twin-spotted wainscot moth lays its eggs in the bur reed, but the larvae later transfer to the common reed as they fatten up and need a thicker stem to tunnel into. In the summer dusk the pale hatched moths float out over the riverside. A specialized fly, *Lipara lucens*, also tunnels into the reed stems, creating noticeable swellings known as 'cigar-galls'. Once the fly has flown, the empty gall provides a winter home for two other reed specialists: a bee, *Hylaeus pectoralis*, and a wasp, *Passaloecus corniger*, whose eggs will hatch in the following spring.

Few species have adapted so closely to their particular rivers as caddisflies, stoneflies, and mayflies, which, in their turn, have been cunningly imitated for bait by generations of fishermen. In the larval stages, caddisflies build themselves cases out of the materials of the river bed. These provide them with camouflage and, depending on the speed of flow, either ballast or a means of transport. The faster the stream, the heavier the material chosen; while those species occupying slow-flowing rivers or ditches construct a case of wood around themselves to help them float to fresh feeding grounds.

The nymphs, or first larval stages of mayflies, are also adapted to very particular conditions. Some nymphs have specially shaped heads and legs, so that when facing the current, they are pushed against stones into which they fit, and which save them from being washed away—certainly a case of going with the elements! The yellow may nymph is best adapted for rough boulders, while the marsh brown nymph fits against smooth stones, upon which its gill-like plates press down, thereby creating a vacuum. While some species are streamlined for fast flows, others are burrowers and bottom-crawlers. The claret dun nymph is at home in slow, peaty streams. It is peat-coloured, and its gills both camouflage it by breaking up its outline and enable it to breathe in still water. The blue-winged olive nymph lives among weeds such as water crowfoot. It is neatly shaped to lodge in close-packed vegetation, from which it can be in close contact with the fast-flowing oxygenated water it requires.

The culmination of all this unseen evolution on the river bed is one of the great phenomena of the

Reed buntings and common reed.

English countryside, once seen, never forgotten. This is the day in the life of the mayfly. Very punctually in mid-May, the nymphs will rise from the bed of the river and hatch through a final nymph stage known to fishermen as 'duns'. Then, when the air is still, the elegant adults, 'the spinners', float upwards in their thousands and perform their mating dance. This is the sight which stays with even the casual observer. The gauzy tides of swarming males, waiting for the females, rise and fall as if on invisible yo-yos. Having mated, grey clouds of females glide to the water, lay their eggs, and die. With all the poignancy of a Shakespeare sonnet, it is over in the space of a summer day, until next spring.

> When I consider everything that grows
> Holds in perfection but a little moment.[4]

The life of a river has nothing to show more resonant of changeless change than the life cycle of the mayfly, a genus known even in the dry language of science as *Ephemera*.

Yet the return of the mayflies is no longer as inevitable as the return of May. They are steadily declining in many rivers, and have vanished from others. Pollution and the removal of riverside hedges have played their part; but above all, dredging and drainage have ironed out the varied bed conditions of gravel and silt to which the larvae of these and many other insects were so minutely adapted.

The otter, sliding up a river like a sleek cat and whistling to its mate under the moon, is truly king of the waters, and the presence of otters on a river system sets the final seal of well-being on its wildlife. The otter has captured popular imagination ever since the classics of Henry Williamson and Gavin Maxwell. It

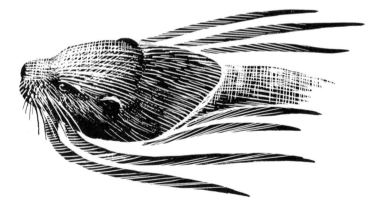

achieved tabloid status in April 1985, when it was on the front page of the *Daily Mirror*'s conservation shock issue; and, as a symbol of wildlife under threat, it is a sure money-maker for such causes as the World Wildlife Fund. The fact that people will give money to save the otter, a nocturnal animal whose presence is detected even by full-time otter survey teams only by its tracks and droppings, is the best answer I know to that mean-spirited and illogical argument: 'What's the use of saving it, if I can't see it?' It was enough simply to know that otters were out there somewhere. Alas, no longer. In 1977 leading conservationists produced a report showing that the otter had declined with disastrous suddenness.[5] Whereas otters were present, even common, throughout the country in the 1950s, they are now abundant only in the extreme north-west of Scotland, leaving core populations in Wales and the West Country, and a dwindling interbred group of individuals in East Anglia. Hunting, disturbance, and pesticide residues had all played their part; but the major culprits were river boards and their successors which scoured the banks of undergrowth in which otters lay up during the day, and felled the mighty riverside trees, such as ash and sycamore, in whose buttress roots otters made their holts. Since 1977 otter hunting has been illegal, and the ban on the pesticide dieldrin is starting to have a beneficial effect. But the many miles of treeless river inhibits recolonization by otters, and even in the 1980s there have been cases of water authority workmen felling known otter holts.

## The Cultural Heritage of Rivers

It is not only wildlife which is at stake. Rivers represent a cultural heritage as well. Through aeons of geological time, rivers, in the wake of glaciers, have helped shape the very structure of our landscape. When man arrived, rivers shaped his history, and men, in turn, began to shape them.

From the time that the tribal Belgae and then the Romans invaded England, rivers dictated the positions of many towns and villages. It is believed that the great bluestones of which Stonehenge is constructed were floated up the Wiltshire Avon. Christianity came to England up a river, when St Augustine and his forty monks travelled to Canterbury up the Kentish Stour, 'singing all the way'. Rivers were also highways of terror. The sleek hull of a Viking boat was specially designed to be shallow enough and narrow enough for use on navigable rivers, along which Norsemen brought fire and the sword. Along the waterways was carried most of the stone required to build our medieval cathedrals. When Whittlesey Mere was drained in the 1850s, a heap of dressed stone was found at the bottom of the lake, evidence of a medieval boating accident. It is thought that the cargo was destined for Crowland Abbey. In the eighteenth and nineteenth centuries the harnessing of water-power for mills and river navigation interlinked with a new system of canals laid the foundation for the Industrial Revolution. Through all this change, rivers have continued to flow; but, for those who have eyes to see, the imprint of each generation remains, varying and concentrating the character of each stream which flows past our homes and our lives.

A good example of this is the water-mill. The Romans introduced water-mills to England, and the 'Domesday Book' records as many as 5,624. All but a hundred of these mill sites can still be accounted for. Up and down rivers and brooks, the remains of some of these and many later water-mills can be seen. In

some cases, magnificent buildings, complete with all their machinery, still stand. Elsewhere there are just clues to former occupancy: foundations, a silted mill-race, or the remnants of a weir, as described by Edward Thomas:

> Only the idle foam
> Of Water falling
> Changelessly calling,
> Where once men had a work place and a home.[6]

All this interference with rivers by man, the building of mill-races and mill-ponds and, in some cases, quite major diversions of watercourses, did not destroy the essential character of rivers. On the contrary, it added to it, and not just in terms of the quality and character of the landscape, but also from the point of view of the wildlife. A tumbling weir creates the localized conditions of an upland brook wherever it crosses a silty lowland stream. Here grow the willow moss and liverworts found again in abundance only towards the sources of a brook. In the highly oxygenated water below a weir swim the little fish not known for nothing as the 'miller's thumb': the flattened head of the fish was often compared with the thumb of the miller, worn it was said from testing the flour. Perhaps the loveliest of upland specialists associated with weirs and mill sites is the grey wagtail—grey in name, but not in appearance, with the flash of his canary-coloured chest. The grey wagtail nests in ledges and crevices of rock upstream, and finds a similar home in the crumbling walls and vertical structures of watermills. Edward Thomas may not have known that he was also making an ecological point when he pinned down so precisely the atmosphere and feel of these places:

> The sun blazed while the thunder yet
> Added a boom:
> A wagtail flickered bright over
> The mill-pond's gloom.[7]

Attractive colonizers of the mill weirs. Grey wagtail and the delicate growth of skullcap.

Some of the wildlife of the water-mill may owe its existence to a rather more conscious decision on the part of some long-dead miller. Growing along old mill-races or near mill-ponds are some of the stoutest and hoariest of pollard willows, which are the glory of any river bank. This is no accident. The willow was put to many uses by millers. An integral part of mill machinery was a simple spring known as the 'miller's willow'. In the most recent mills it was made of steel; but more commonly it was a piece of springy willow wood collected from a convenient pollard. Willow was also used for eel traps, and the fact of a river being carefully directed to drive a water-wheel has always made water-mills very convenient places to catch eels. The Luttrell Psalter of 1338 illustrates a water-mill complete with eel traps which look very much as if they have been made out of pliant willow stems. Many medieval millers paid their rent to the lord of the manor in eels; and when the water-mill in the centre of Stafford was pulled down after the last war, the laconic miller expressed as his only regret: 'I shall miss the eels.'

More recently, eel traps have been built into the systems of weirs and sluices of water-mills. Yet these structures, too, add variety and local character to rivers. The joints in a timber lock-gate or the eroding mortar of a sluice often provide congenial conditions for gipsywort or skullcap, with its clear blue flowers. Both these rather delicately proportioned plants have more difficulty competing with other vigorous vegetation on the open river bank than they do in the neat crevices which man has provided for them. The structures themselves were often built of local materials. Before the advent of railways, it was cheaper to do this than to import materials from far afield. Later, bricks were commonly imported, but even these, including the splendid 'blues' of the industrial Midlands, added their individual stamp to river landscapes. Nevertheless, it is surprising how often local builders simply took advantage of what was at hand. In 1985 a land-drainage scheme was carried out on the river Erewash near Eastwood in Nottinghamshire. This is the river which flows as a constant theme through the novels of D. H. Lawrence. Just opposite the point where the brook which runs past Lawrence's old home joins the Erewash itself stands a mill sluice. In the normal course of events this would have been 'tidied up'—wiped out as part of the scheme and conveniently buried. But in this case the enlightened engineer was concerned to do precisely the opposite, and as the digger pulled back the rubble, he exposed an area of wall which looked for a moment as if it was made entirely of shiny cream bones. This turned out to be pottery waste, a surviving memorial to a now vanished china factory in the local town. There it now stands, its cracks colonized by ferns and wormwood—nothing especially beautiful in itself, but a piece of history and natural history, with literary associations thrown in.

Builders in the wetlands did more than that however. They built their homes out of the materials of the river bank itself. Whereas in most parts of England, thatched roofs were made of locally available straw, in the lowlands, especially in East Anglia, the common reed was used—the same plant which provides a

home for a whole hierarchy of warblers, moths, and bees. And very good thatch it made. It is still often known as Norfolk reed; and in contrast to ordinary straw thatch, which has a lifetime of thirty years at most, a well-laid thatch of Norfolk reed may last as long as eighty years. In the wettest wetlands of all grows an even tougher thatching material, one of our most ancient natural crops: the giant saw sedge *Cladium mariscus*. This, one of the dominant plants of the undrained Fens, can still be seen on the roofs of some of the houses between Ely and Newmarket. In some cases this most durable, but increasingly unobtainable, material is used as a capping ridge to a thatch of Norfolk reed. In other places, where even the Norfolk reed is scarce, the reed is used as a capping ridge to straw thatch. The final flourish in this humanizing of the natural landscape is given by an individual, as the thatcher in Ronald Blythe's *Akenfield* explains: 'We all have our own pattern; it is our signature you might say. A thatcher can look at a roof and tell you who thatched it by the pattern.'[8]

Elsewhere man has modified the landscape less consciously, by providing congenial conditions for the river water crowfoot. The water crowfoots are a particularly lovely group of river plants. To the layman they all look pretty similar: crisp emerald weed buoyed up in the stream and then, in July, a snow in summer of glistening white flowers, which spill over the water in a way that seems to spell out the brief abundance of midsummer. The nine or so different forms and species to be found in England are all specialists. The pond water crowfoot has broad-lobed leaves which float on the still surface, in addition to dissected underwater foliage. At the other end of the scale are crowfoots of fast waters, whose only leaves are bunches of slender threads which run with the current—the 'crow's feet'. The river water crowfoot, *Ranunculus fluitans*, grows in gravel, and dies if it becomes too covered up by silt. Studies made on the river Lugg in Herefordshire show that the rubbly remains of fords or bridges which collapsed long ago have created a gravelly river bed which encourages crowfoot. The Lugg is in general a silty river, although it has natural riffles of gravel where the crowfoot also occurs. Elsewhere, however, known historical fords or bridge sites can be picked out every high summer when the crowfoot blooms—literally, living history.

The same effects are visible in the River Wye in Hereford itself where the site of the early ford below the Bishop's Palace is picked out in white in July when the water crowfoot is in flower. There is even a gap in the centre with no flowers where the material marking the ford was probably removed at a later date to allow passage for the boats, leaving a silty bottom to the river in that place unsuitable for the growth of water crowfoot.[9]

Our predecessors' efforts to farm and drain the land can also be 'read', again in white and green, at a very different but particular moment of the year. If you stand on a hill in winter when there is a thaw or look down from a motorway, especially in the Midlands, you can hardly miss the pattern of long parallel bands of snow, which are always the last to go from the hollows of the old ridge and

*Top.* A pollard willow beside a water-mill. *Bottom.* 'The miller's willow', a springy piece of timber, still gathered from the convenient pollard, for use as an essential part of the mill machinery. Charlecote, Warwickshire.

furrow. In summer, too, you can read the ridge and furrow in an even more attractive way, although this is possible only on those few fields which have not been 'improved' to an all-over green of fertilized rye-grass. In such meadows there is a specialized flora for the damp furrow bottoms, subtly distinct from the flowers and grasses of the dry ridge crowns. The bulbous buttercup (*Ranunculus bulbosus*) prefers the tops, while the creeping buttercup (*R. repens*) favours the hollows. Ridge and furrow were formed any time between the early Middle Ages and the nineteenth century, as a result of ploughing up and down fields in parallel lines. This is generally thought to have been done deliberately, to improve drainage in the days before clay and now plastic field drains. In fields next to a river, the furrows are noticeably at right angles to the stream, although in some other places, where no obvious drainage benefit was gained, ridge and furrow seem to have been simply a by-product of the normal way of ploughing. If you look carefully, you may see some ridge and furrow which lie in a reverse-s pattern, a result of the logistics facing the medieval ploughman, who had to manœuvre eight oxen up and down a field. The Tudor ploughman had better-bred and better-fed beasts; so he required fewer of them to pull a plough, and was able to drive a straighter furrow.

These are the monuments to generations of individual farmers ploughing and draining their fields. When, in the seventeenth century, the first real drainage engineers arrived, equipped with a literally world-changing technology, they too left their monument, in the shape of an entirely new landscape: the Fens. Despite their new-found power, however, they did not succeed in totally obliterating what had been before. Meandering across the geometrical landscape of straight ditches and square fields which the new engineers and their successors created, the winding courses of the original pre-drainage creeks and rivers can be seen especially clearly from the air. As they flowed on their—as it turned out—not so eternal journey from the hinterland of Ely and Cambridge to the Wash, these rivers deposited silt along their beds; and now, although the rivers vanished centuries ago, this silt stands out, a startling pale-fawn colour, as it snakes across the adjacent black peat of the Fens. Thus you can still pick out with ease the route of the ancient Wellestream, up which, in the fourteenth century, came cargoes of cloth from the Low Countries and silks from Italy, not to mention news of the dawning Renaissance, bound for Cambridge and beyond. What is even more astonishing is that you can see the Wellestream more easily with every passing year. This is because the adjacent peat, as it is drained and dried out, wastes away by a process of oxidation on exposure to the atmosphere. Hence the old silt river beds, known in the Fens as 'roddons', are rising steadily as ridges above the ever-falling peat. They are landscape ghosts; but instead of fading away, they sharpen ever more clearly into focus.

Rivers have always been boundaries, as well as route-ways. They dictate the shape of many land-holdings, parishes, counties, and even parts of the Welsh and Scottish borders, as is well known to the poor river engineer who has to

Snow picks out the ancient ridge and furrow, which is overlain by enclosure hedges. Warwickshire.

negotiate with different landowners, not to mention councils, on opposing banks as he tries to promote his scheme. With boundaries go hedges, and some of those still remaining have been part of the farmed landscape since Saxon, or possibly even earlier, times. Boundary hedges tend to be the oldest hedges, and a number of these are found bordering brooks and streams. Techniques developed by botanists and historians have shown that it is possible to assess the age of a hedge by the number of shrub species it contains.[10] So the hedges which most interest the historian are those which most fascinate the botanist. To the layman they are also arguably the most beautiful, with all the tangled richness and variety of oak, ash, buckthorn, elder, and wild rose. In general, a hedge will contain in a 30-yard length one shrub species for every century of its life. This is not an immutable law, but the correlation has been sufficiently demonstrated to be a valuable guide.

These are the relationships which print the character on the landscape. It is the same with the various damselflies in their clay and gravel streams, the water-mill with its grey willows and grey wagtails, and the crowfoot crowning the ford and the curve in ridge and furrow. This is what the Englishness of England is made of, and it is this sense of place and the particular which we are currently in danger of losing, just as we come to understand it more clearly. It is far more important than any particular beetle or bird or ancient monument. The long evolution of the natural system, further evolved by man's practical management, has given brooks and rivers individual characters as distinctive as their names, names which invest the Ordnance Survey map with an idiosyncratic quality ranging from poetry to comedy: the Windrush, the Swift, the Wagtail Brook, the Mad Brook, the Hell Brook, the Piddle, the Wriggle. . . .

Take another look at *The Haywain*. Constable was a miller's son, and probably knew all there was to know about mills such as Flatford. He was concerned here above all to paint a *working* landscape, not just a decorative one; and the empty wain is fording the stream to collect more hay from the labourers in the distant field. It was probably there for another reason too. George Sturt, who wrote about his practical experience of running a wheelwright's shop at the end of last century, describes the constant problem of shrinkage of timber wheelstocks in high summer:

Men who used carts knew something about the advantage of a little moisture for tightening wheels. Not for the horse's sake alone was it that carters would drive through a roadside pond, or choose to ford a stream rather than go over a bridge beside the ford. The wheels were better for the wetting.[11]

Sturt and others record how wagons for the Sussex clay country were designed with broad wheels, while those for downland chalk had narrow ones. Taking individual orders from his customers, the wheelwright built each cart for the particular conditions of a particular farm. This is taking a sense of the particular about as far as you can go: man evolving as harmoniously with his landscape as

The ghost of a long-vanished river winds across the level geometry of the Fens. The silt bed of the old river now stands out against the adjacent black peat. Photo: Cambridge University Collection of Air Photographs.

Cartwheels soaking in the river in order to tighten them. Constable's famous haywain was part of a working landscape, not simply a picturesque one. National Gallery.

the mayfly nymph evolved in harmony with its stream. No wonder farmers tell us that they are the creators and guardians of the English countryside.

Now, for the first time in the long history of settling our islands, the guardians have become the destroyers. The landscape which generation after generation created is like a classroom blackboard at the end of a day, on which each lesson has been written without entirely erasing the previous one. Medieval ridge and furrow are overlain by a grid of more recent hedgerows. The winding Wellestream can be glimpsed through the level geometry of the Fens. In the space of a generation, we have set about wiping the blackboard clean.

## The Impact of River Management

Rivers and streams have been straightened and evened out as never before. This work has been carried out by water authorities, internal drainage boards, and many councils, in part to reduce flooding of roads and houses, but largely to increase farm yields. In the years between 1971 and 1980, an annual average of 207,217 acres (83,858 hectares) was estimated to have been drained, of which 'new' drainage of wetlands comprised around 20,000 acres per year.[12] One straightened stream begins to resemble another. River organisms' ability to survive the disruption of floods was never evolved to withstand this kind of onslaught. Repeated dredging has removed the weedy margins upon which dragonflies depend. In the last twenty-five years, four dragonfly species have become extinct in England, while many others have shown a marked decline. Reduction of such insect populations will in turn reduce the fish population, which depends on a diversified, rather than a straightened and uniform, channel. With the loss of the fish, we can expect to lose the electric-blue flash of the kingfisher. The dredgings are put in the bottoms of furrows or are used to fill in ponds. With the virtual disappearance of the farm pond, frog populations in some parts of England declined drastically between the 1950s and the 1970s.

Many streams have been stripped of their ancient boundary trees, and the knock-on effect of the drainage schemes has been to encourage farmers to turn their farms into prairies. Between 1946 and 1963, around 85,000 miles of hedges were grubbed out.[13] The lowering of water-levels to allow ploughing of damp pasture has removed the nesting habitat of many birds which we used to take for granted, such as snipe, lapwing, and redshank. A survey of Oxfordshire in 1982[14] found only 15 breeding pairs of redshank, compared to 112 pairs in a comparable survey in 1939. Studies have shown that the water-vole, 'Ratty' of *The Wind in the Willows,* has become scarce in many areas, due to intensive management of river banks.[15] Indeed, Ratty's emotional, if not his ecological, headquarters, the river Pang between Reading and Newbury, where Kenneth Grahame wrote his classic, has been subjected to a notoriously insensitive drainage scheme. In some counties, such as Staffordshire, such a previously common wetland plant as reed has become a rare sight. The remains of

Water-vole.

water-mills have been consistently removed; the weirs which hold up river levels have been dismantled; and mill-ponds and mill-races have been filled with the dredgings. On the Worcestershire Stour, ten mill sites have been removed in as many years. The ancient fords which hold up the water are dredged out, along with their water crowfoot. In the late 1970s, the Anglo-Saxon 'stretford' of Stratford-upon-Avon was taken out.

The inexorable desiccation of the Fens, the Somerset Levels, and the Lancashire mosses is exposing remarkable artefacts of early man, long preserved in the wet peat. But in the absence of an effective liaison between drainage men and archaeologists, there is a danger that such remarkable finds as the Bronze Age settlement built on a timber 'island' recently rescued at Flag Fen near Peterborough could be broken up by diggers or else left to crumble on exposure to the atmosphere.[16]

Major wetlands are obvious victims of drainage. Casualties since the War have included the river Idle washlands in Nottinghamshire and large parts of Romney Marsh, Otmoor, and the Lancashire mosses. Major debates have been held since the late 1970s over the future of wetlands in Sussex, Somerset, Yorkshire, and East Anglia. Wetlands under threat have included a variety of landscapes: swamps of tall reed or reed sweet-grass; marshes of rush and sedge, which sometimes develop into scrub of willow and bog myrtle; fens, whose lush vegetation is nourished by alkaline groundwater, and which range from open pools, often the remains of peat cutting, to grazed beds of meadowsweet and iris, grading in turn to the wet woodlands known as alder 'carr'; mires, such as the mosses of the North-West, whose deep peatlands support sphagnum moss and heather, scattered with glades of birch, the favourite haunt of nightjars. Finally, there are flowery haymeadows and damp pastures, intersected by dykes patrolled by dragonflies in summer and all submerged in winter by the silver flood, which draws in dark clouds of wildfowl and companies of wild white swans. The destruction of such places in our time has been startling. In 1983 the chief scientist of the Nature Conservancy Council, the Government's watchdog on conservation, produced a definitive report on the destruction of habitat in Great Britain since 1949.[17] Among the casualties wholly or partly attributable to drainage were 97 per cent of herb-rich haymeadows, 50 per cent of lowland fens, and 60 per cent of lowland raised mires, all lost in the space of a generation. In the mid-1980s the Norfolk Broads were losing an average of 1,500 acres per year.

All this has happened because enormous sums of public money available for drainage schemes since the War have combined with a revolution in technology which we have not yet fully come to terms with. I will always remember standing one early spring day near the river Severn, arguing with a drainage officer who had previously maintained watercourses now due to be reshaped as part of an expensive new scheme. This man had as little sympathy for the environment as a pike might have kind feelings towards a minnow. Yet the point which he rightly argued was that for the previous thirty years he had personally controlled the maintenance of the ditches and hedges of this parish, so why should he consider newfangled ideas about nature conservation being built into any proposed scheme now? I looked at what he had created. In spite of his indifference, it was exquisite: the ditch banks were creamy with cowslips and lilac with cuckoo flower. Chiff-chaffs were arriving to nest in the spangled scrub of blackthorn. There was a badger set under some old pear-trees. I then asked him how he managed the place. It turned out that he had a small dredger, an even smaller budget, and a very primitive brand of mowing machine. With these he had efficiently kept water flowing through his ditches. I lost the argument that day. Within a week the big machines had moved in, and that corner of his parish could have been one of thousands of others in modern England—any place, anywhere.

The lesson was clear. Of course, the countryside must continue to be a working landscape; but if most people's definition of a river as something more than just a drain is valid, then that broad definition must be *consciously* built into the brief of those who wield this mighty technology of the JCB, the Hymac, and the Swamp-dozer. Only then can we guide the evolution of the countryside within legitimately broad terms of reference and continue the age-old process of civilizing the rivers. And why not? The big machines are only powered by the ratepayers' money, and that woman who threatened to tie herself to a willow tree represents thousands upon thousands of ratepayers who share her (and Constable's) convictions about the essential nature of a river.

Over the last half-dozen years there has been a quiet revolution in the water industry as this simple realization has dawned upon engineers, farmers, digger-drivers, and even the legislators in Westminster. Of course, there are still places where the old-style canalizing approach to river management is being pushed through; and the conflict of values which underlies this whole issue raises a number of questions which are not easy to answer. Nobody can seriously suggest that we turn back the clock entirely and return to the world of Constable's *Haywain*, where there was a good deal of misery and hunger amidst all that beauty. We admire and cherish an environment which we also depend on for food. We may reduce the dredging of rivers, but if we stop it altogether, floods will return to overwhelm us. We are therefore committed to continue managing rivers, as we are to managing every square mile of the English countryside. It is the *way* we do so which counts.

Before work began.                    The machines move in.

The fact that rivers are such a symbol of endurance and of changeless change is what makes their management a touchstone for the whole issue of our relationship with the natural world. It is therefore a moving thought that the rivermen were among the first people in the modern countryside business to begin to stop and think a little harder about what they were actually doing. To adapt the slogan 'Put the Great back in Britain', some of them have begun to put the river back into river management. It remains to be seen whether at this eleventh hour for the English countryside, those other giants, the forestry and the agriculture industries, are also prepared to take seriously a wider frame of reference. If they are, we will at last be able to see the countryside put back

Three stages in the destruction of a river.

The disastrous results of traditional river engineering.

into countryside management. Such an achievement depends upon two things for which the English have always had a special genius: a sense of place and a sense of compromise. What river engineers have begun to do is to rediscover their roots, and these, as we shall see, go back a very long way.

# 2

# *The Fear of the Flood*
## *Traditional Attitudes to Wetlands*

IN the beginning, the waters covered the earth. The first thing you would notice about the landscape if you were to travel back in time was how wet it was. In prehistoric times rivers and streams ran unbridled over their flood plains, and most low ground consisted of marshes, fens, and very wet woodland. Well into modern times the major wetlands of England remained undrained: the Vale of York, the fens around the Humber, the Essex marshes, the Lancashire mosses, Romney Marsh, the Severn lowlands, the Somerset Levels, and, above all, the 'Great Level' of the Fens. Surveying these now prosaically productive acres of beet and potato, it is as hard to imagine their undrained state as if one were trying to conjure up some fabulous landscape lost beyond recall. Charles Kingsley, writing of the final destruction of the Fens, was perhaps the first to regret the loss of what must have been one of the finest natural systems in Europe. He describes immense tracts of pale reed and dark-green alder stretching from Cambridge to Peterborough, from King's Lynn to the foot of the Lincolnshire wolds, where:

high overhead hung, motionless, hawk beyond hawk, buzzard beyond buzzard, kite beyond kite, as far as eye could see. Far off, upon the silver mere, would rise a puff of smoke from a punt, invisible from its flatness and its white paint. Then down the wind came the boom of the great stanchion-gun; and after that sound another sound, louder as it neared; a cry as of all the bells of Cambridge, and all the hounds of Cottesmore; and overhead rushed and whirled the skein of terrified wild-fowl, screaming, piping, clacking, croaking, filling the air with the hoarse rattle of their wings, while clear above all sounded the wild whistle of the curlew, and the trumpet note of the great wild swan.

They are all gone now. . . . Ah, well, at least we shall have wheat and mutton instead, and no more typhus and ague; and, it is to be hoped, no more brandy-drinking and opium-eating; and children will live and not die.[1]

There speaks the nineteenth century: all gone, but in a good cause. It is only our more ambivalent age, engaged in the very last mopping-up of the great wet waste which challenged our ancestors, which has begun to question whether the price of progress has been too high. We can see more clearly now that the ultimate

Major wetlands present in the seventeenth and eighteenth centuries.

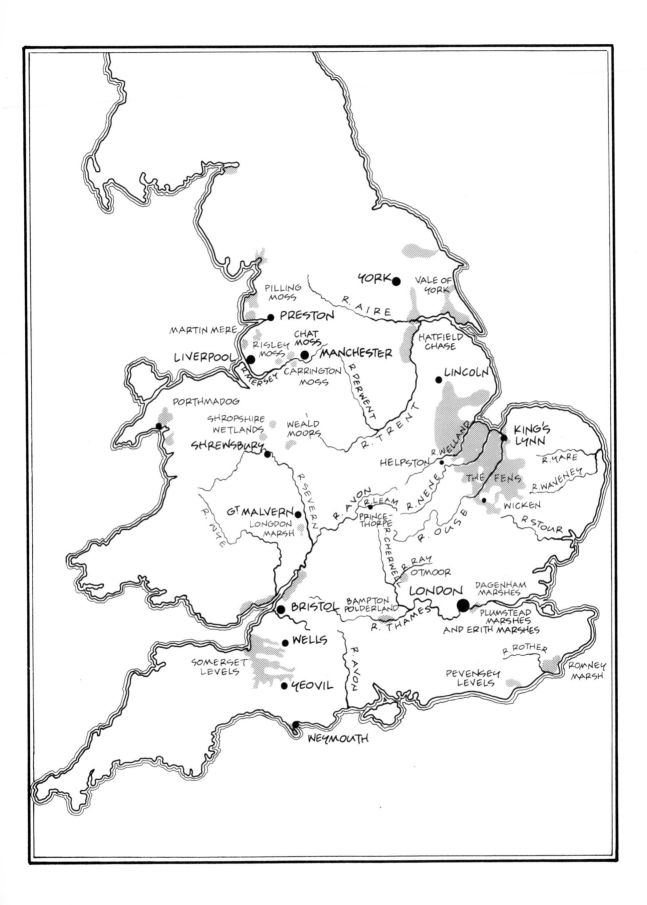

YORK •   VALE OF YORK

PILLING MOSS

R. AIRE

MARTIN MERE   PRESTON •

CHAT MOSS

RISLEY MOSS   HATFIELD CHASE

LIVERPOOL •   MANCHESTER •

R. MERSEY   CARRINGTON MOSS   LINCOLN •

R. DERWENT

PORTHMADOG

SHROPSHIRE WETLANDS   WEALD MOORS

R. TRENT

SHREWSBURY •   KING'S LYNN

R. SEVERN   R. YARE

HELPSTON •   R. WELLAND

R. WYE   Gt MALVERN •   THE FENS   R. WAVENEY

LONGDON MARSH   R. AVON   R. LEAM   R. NENE   WICKEN •

PRINCE-THORPE   R. OUSE   R. STOUR

R. CHERWELL

R. RAY

OTMOOR   DAGENHAM MARSHES

LONDON •

BRISTOL •   BAMPTON POLDERLAND   PLUMSTEAD MARSHES AND ERITH MARSHES

R. THAMES

WELLS •   R. AVON   R. ROTHER

SOMERSET LEVELS   PEVENSEY LEVELS   ROMNEY MARSH

YEOVIL •

WEYMOUTH •

end of such a process—the flood entirely tamed—is both impossible and undesirable. The long, and still unfolding, history of land drainage contains much that is progress in the best sense: the combating of diseases and terrifying floods and the production of food to sustain a growing population. But it is also a saga of human avarice and the abuse of power.

Quite apart from the major wetlands, every valley bottom below a certain contour line must have been soggy and at times impassable. Quite how wet any particular locality once was can often be guessed through detective work involving an enjoyable study of plants, place-names, and local history. At Henley-in-Arden in Warwickshire, two churches face each other across the little river Alne. Their presence is explained in an appeal of 1548 to retain both churches: 'The town of Henley is severed from the Parish Church with a brook which in winter so riseth that none may pass over it without danger of perishing.' Near by, in the tell-tale peat, are stands of meadowsweet and sedge, the last remnants of Henley's ancient marsh, now happily salvaged as an oasis within a new housing estate.

In other places, canals have provided a damp lifeline for plants surviving from much earlier wetlands. In the 1950s the redoubtable Eva Crackles, a Yorkshire teacher, was gathering grasses at the point where the Leven canal crosses the site of an ancient lake, now long vanished, but clearly just surviving when the canal was cut in 1802. For there, just at the point where canal and lake site coincide, she found one of the few recorded colonies of the narrow small reed, tenaciously clinging to the mud. Newport in Shropshire was 'new' in the twelfth century when it was granted a charter by Henry I, who required that it supply fish to the royal household from its medieval fish-pond. This pond survived just long enough to be incorporated into the Shropshire Union Canal in 1833, complete with an unusual wealth of water plants, which have earned the canal basin the status of Site of Special Scientific Interest in 1986. At Gailey in Staffordshire, an unusually rich flora emerged in gravel workings abandoned in the 1960s, relics from the fen from which the place took its name, which comes from the Anglo-Saxon 'gagol leah', the clearing in the gale, or bog myrtle, that aromatic wetland shrub which gives an extra tang to home-made gin.[2]

Names on the map tell us a great deal about the ancient undrained landscape, and none is more telling than the presence on the Ordnance Survey Map of the lowlands of the word 'moor': Morton or Moortown; Sedgemoor; Otmoor; Moorgate, the gate in London's city wall which opened on to Moorfields, the marsh which William Dugdale, in his seventeenth-century classic on drainage, describes as a favourite resort of Londoners for skating.[3] The men of the Somerset Levels paid a tithe called 'moor-penny'; their cattle suffered from a disease called 'moor-evil'; and in every pond and damp corner you will still see the jerking movements of the moorhen. The *Oxford English Dictionary* gives as its first meaning for 'moor' 'uncultivated ground covered with heather'. To most of us, reared on *Wuthering Heights*, that is what a moor means. But to our ancestors, living when the hills were less thoroughly cleared and the lowlands were more

universally wet, a moor was something more terrifying: a morass. The word 'mor' first occurs in
Saxon accounts of King Alfred hiding in his wetland
fastness in Somerset, and most evocatively of all
in our national epic poem *Beowulf*. The hero,
Beowulf, does battle with Grendel and Grendel's
mother, two enormous monsters which haunt the
swamps—'moras' in Anglo-Saxon—from which
they emerge to wreak havoc before returning to a
mere in the very heart of the fen: 'The lake which
they inhabit lies not many miles from here, overhung with groves of rime-crusted
trees whose thick roots darken the water.'

Moorhen.

## Hostile Wetlands

This description, at the very start of our literature, sets the tone for accounts of
wetlands, which through the ages have had a consistently bad press. When in
the eighth century the Saxon saint Guthlac penetrated the heart of the Fens to
found Crowland Abbey, he was described by the monk Felix of Crowland as
encountering demons in the wilderness, which 'came with such immoderate
noises and immense horror, that it seemed to him that all between heaven and
earth resounded with their dreadful cries'. They bound Guthlac 'in all his
limbs . . . and brought him to the black fen, and threw and sank him in the
muddy waters'.[4]

With the passage of time, demons are about the only form of unpleasantness
not recorded in accounts of the wetlands. William Lambarde, Elizabeth I's
archivist, described Romney Marsh in 1576 as 'evil in winter, grievous in summer
and never good'.[5] In 1629 the Fens were vilified thus: 'The Air nebulous, grosse
and full of rotten harres; the water putred and muddy, yea full of loathsome
vermine; the earth spuing, unfast and boggie.'[6] ('Harres' were noxious gases.)
Samuel Pepys, visiting his relations at Wisbech thirty-five years later, was equally
unimpressed as he passed through 'most sad fennes, all the way observing the
sad life which the people of the place—which if they were born there, they do
call the Breedlings of the place—do live, sometimes rowing from one spot to
another and then wadeing'.[7] For travellers such places provided a multitude of
hazards. At best they involved a detour. At worst there was the danger—horror
of horrors!—of falling in. The intrepid lady traveller Celia Fiennes had a near
miss when her horse was nearly sucked into a dyke near Ely in 1698; and in the
same year she took care to avoid Martin Mere in Lancashire, 'that as the proverb
sayes has parted many a man and his mare indeed'.[8]

The fate awaiting someone pitched from his horse in such a place might be
blood-poisoning, 'being dreadfully venom'd by rolling in slake', as William Hall
put it in his nineteenth-century fen doggerel.[9] Worse still, one might be swallowed

To outsiders, wetlands appeared hostile fastnesses, associated only with floods and disease.

for ever in the morass. Daniel Defoe wrote of Chat Moss, near Manchester, as 'being too terrible to contemplate for it will bear neither man nor beast'.[10]

Getting lost was another likelihood, unless, as at Longdon Marsh in Worcestershire, the traveller was able to pay a guide to show him across. On the swampy willow scrub of the Wealdmoors in Shropshire, the local rector described in 1673 how 'the inhabitants commonly hang'd bells about the necks of their cows that they might the more easily find them'.[11] Otmoor was notorious as a place in which to get lost, and verses celebrate how the curfew rung on winter nights from Charlton church guided travellers out of the intractable moor. Fog, the one element which no drainer can ever quite banish from the marshes, still rolls out over Otmoor. A farmer's wife giving evidence at the Otmoor M40 inquiry in 1983 described how she had once become completely lost in one of her own fields while counting sheep. Daniel Defoe describes the Fens shrouded in fog, through which nothing could be seen 'but now and then the lanthorn or cupola of Ely Minster'.[12] To further terrify lost, wandering travellers, igniting marsh gas created the alarming phenomena, still not fully understood by scientists, known as will-o'-the-wisps, jack-o'-lanterns, or corpse-candles.[13]

'Infect her beauty, / You fen-sucked fogs,' inveighed Shakespeare's King Lear against his daughter. Our ancestors associated wetlands with disease. They had good reason. As late as 1827, travellers were 'fearful of entering the fens of Cambridgeshire lest the Marsh Miasma should shorten their lives'.[14] On the Somerset Levels, inundated by heavy floods in 1872 and 1873, a report described how 'Ague set in early in the spring and is now very prevalent . . . among the poorer families who are badly fed and clothed.'[15] 'Ague' was malaria, meaning literally 'bad air', the marshy miasma which, until the discovery of the malarial mosquito in 1880, was believed to be the main cause of the disease. The malarial mosquito, *Plasmodium vivax*, breeds far north into Europe, and although less virulent than some tropical species, it was still responsible for many deaths. Malaria was endemic in the English wetlands. 'As bad as an Essex Ague' was a common expression;[16] and in the 1870s the garrison at Tilbury Fort was changed every six months because of the prevalence of malaria. The Thames marshes ensured that the ague was carried into the courts of kings, who were less resistant to it than the hardy fenmen. James I was declared by his contemporaries to have died of it, and his victim Sir Walter Ralegh, awaiting execution in the Tower, prayed that he would not be seized by a fit of ague on the scaffold, lest his enemies should proclaim that he had met his death shivering with fear. The terror, if not the actuality, of the disease has survived into our own time. In the early 1970s Strood District Council was spraying the dykes in the north Kent marshes with DDT as a precaution against malaria.

As towns grew larger, they began to pollute the adjacent marshes and valley bottoms, which in turn developed ominous reputations for disease. Bubonic plague is not directly associated with water, but the rats which carried it arrived by boat at riverside wharves. The Great Plague of London is said to have broken out in 1665 in a marshy district known as the Seven Dials, and it was especially prevalent along the old river Fleet. In the nineteenth century the stagnant waters of cities were haunted by the shadow of cholera. It is no accident that many slums were built on marshes: Mosside in Manchester, the Bogside in Londonderry, and much of the East End of London, where the suffix 'ey' to many of the place-names tell us that they were islands in Saxon times: Hackney, Stepney, and, most notorious of all, Bermondsey, where, in the 1850s, the river Neckinger, 'the colour of strong green tea', flowed round Jacob's Island, which was used by Dickens as a setting for *Oliver Twist*, and was described by him as 'the filthiest, the strangest, the most extraordinary of the many localities that are hidden in London'. Social reformers were not slow to describe the horrors of such places. Friedrich Engels singled out the river Aire in Leeds and the Irk in Manchester for special mention: 'In dry weather, a long string of the most disgusting, blackish-green slime pools are left standing on this bank, from the depth of which bubbles of miasmatic gas constantly arise and give forth a stench unendurable.'[17] A hundred years later George Orwell described the stagnant pools of the Ince flashes at Wigan as 'covered with ice the colour of raw umber . . . nothing existed except smoke, shale, ice, mud, ashes and foul water.'[18]

Add to this the limitations which wetlands impose on farming—short grazing seasons, foot-rot in sheep, suppression of root growth in the damp soil, and the hazards of high water for cereal crops, not to mention the terror of a flood—and it is enough to make one want to rush out and drain all remaining wetlands on sight. Certainly, it is easy to understand why drainage was regarded as a major manifestation of progress. But there is another side to the story. It is a curious fact that the poor benighted people who were unfortunate enough to live in the rural wetlands did not seem to share the prejudices of their visitors at all. Celia Fiennes, in high disgust at finding 'froggs and slow-worms and snailes in my roome' when lodging in Ely, had the honesty to qualify her personal dislike for the place, which 'must needs be very unhealthy, *tho' the natives say much to the contrary* which proceeds from custom and use'.[19]

### The Harvests of the Wetlands

It really was exasperating to observe how the natives seemed to like their marshes. William Elstobb found the eighteenth-century fenmen content with 'uncomfortable accommodations';[20] and Vancouver wrote of Burwell in 1794: 'Any attempt in contemplation for the better drainage of this fen is considered hostile to the true interests of these deluded people.'[21]

Back in 1646, one of the few articulate defenders of such deluded people maintained that those who would undertake drainage 'have always vilified the Fens, and have misinformed many Parliament men, that all the Fens is a mere quagmire . . . of little or no value: but those which live in the Fens, and are neighbours to it, know the contrary.' The anonymous author of *The Anti-Projector* proceeded to list the bounty of the fen. There were horses, cattle, fodder, sheep, osiers, and reed; and 'Lastly, we have many thousand cottagers, which live on our Fens, which must otherwise go a-begging.'[22]

Such people knew how to make the watery wilderness yield up its riches. Pre-eminent among the benefits was summer grazing. The very word 'Somerset' (*Sumorsaetan*) is Anglo-Saxon, possibly meaning 'summer dwellers', men who came down to graze the levels in summer-time. It is thought that those who occupied the Malvern hill forts may have herded their cattle down the Worcestershire drove-ways to pasture them on Longdon Marsh in the summers before the Roman conquest. Shortage of grass in high summer was a continual problem in the open-field system of the Middle Ages. No such lack of lush pasture afflicted the Fens, especially in the silt belt, where medieval prosperity is commemorated by mighty churches and confirmed by historians' research into medieval and sixteenth-century tax returns. It was the flood itself which often ensured the rich

The inhabitants of the wetlands depended upon them for their survival. Economic resources of the Fens—pasture, wildfowl, and domestic geese—are here illustrated on the village sign of Cowbit in Lincolnshire.

grazing. The commons of the Isle of Axholme lay under water from around Martinmas (11 November) until May Day. As the inhabitants were to inform the men who set out to drain these wetlands in the seventeenth century, this flood brought with it 'a thick fatt water'. After drainage had removed this regular winter flood, the people were left with 'thin hungry starving water', which rendered the land incapable of supporting the large grazing herds which it had formerly sustained.[23] Even now, local farmers around Axholme regard the 'warped land' which was deliberately flooded with silt as the best land in the region. Cobbett declared that the marshes of South Holderness in the East Riding, together with the Fens, were the richest land in England.[24] What cheeses must such land have produced! A cheese resembling Camembert was the glory of Cottenham in Cambridgeshire, where records for cheese making go back to as early as 1280; and production ceased only in the mid-nineteenth century with the enclosure of the common fen.[25] Domestic geese, herded on the Fens and the marshes, kept the rest of the country supplied with quill-pens. On the wetland commons of western France, geese are still kept, and furnish a nice line in duvets.

In the wettest and wildest parts of the marshes, fishing and fowling replaced more organized farming. The terse Latin of a Cambridgeshire assize role records how, in the fourteenth century, a boy went out on stilts after birds' eggs and was drowned in the heart of the fen.[26] With the passage of time, such perilous subsistence gave way to more profitable wildfowling. Birds were netted and exported to London. A check-list that would make a modern bird-watcher salivate was served up on the Elizabethan menu: 'the food of heroes, fit for the palates of the great',[27] as Camden describes pewits, godwits, knot, and dotterel. Fish were also exported. Daniel Defoe saw fish transported live from the Fenland to London 'in great butts fill'd with water in waggons as the carriers draw other goods'. Islip eels from Otmoor supplied the Ship Inn at Greenwich. Eels, speared through their gills on an eel stick, had long been standard rent in the Fens and Somerset Levels. 'Ely' itself means the district of eels. The method of catching eels with a glaive or trident lasted just long enough in the Fens to be recorded on an early documentary film.

Eel and eel glaive.

Many wild plants of the wetlands were also harvested. Until the mid-nineteenth century, basket-makers were actively cutting willow at Beckley on Otmoor, a place which also sent water-lilies to Covent Garden. Purple moor grass, which forms the pale-fawn undercarpet of the scattered birch woodlands of such wetlands as the Lancashire mosses and Hatfield Chase was popular for cattle bedding. From Burwell Fen, sedge was sent out by boat for the purpose of drying malt, and Cambridge imported fen sedge as kindling. Bedmakers in Cambridge colleges were issued with stout gloves to protect their hands from the sharp sedge as they lit the fires in undergraduates' rooms. Clogs were made from alder, and reed was used for thatch.

Oak for building the ships of the British navy has always been famous as a resource essential for our national survival. A commonly grown crop of the wetlands was almost as important. Hemp, from which the word 'hempenspun', or 'homespun', originates, is a fibre crop still grown in Russia and the Third World. In England it provided sails and cables for the fleet; and for this reason, legislation going back to the reign of Henry VIII required that a small proportion of land be set aside for its production. This is a far cry from the laws which now pertain to this crop, more familiar today under its Latin name of *Cannabis sativa*. It flourishes best in deep moist ground, and Michael Drayton described south Lincolnshire as 'hemp-bearing Holland's Fen'.[28] On the Isle of Axholme, where little wool was produced, hemp was the basis for a spinning and weaving industry, which provided a useful sideline for the average peasant, and a basic livelihood for the poor.

Peat was always a major wetland commodity. Rights of turf cutting, known as 'turbary', existed in Somerset and the Fens in the Middle Ages. Peat cutting became a major industry in the Lancashire mosses in the nineteenth century, and on maps of Wicken Fen and Hatfield Chase, you will find 'Poor Piece', which was where the local cottagers could cut peat for themselves, subject to regulations prescribing a limited season for peat cutting and insisting that a man may extract as much peat as he can, provided he does not employ assistance. Unlimited plundering of timber was similarly controlled. In 1337 a certain Robert Gyan was submitted to a brutal penance by the dean of Wells for carrying away 'a great number of alder' from Stan Moor in the Somerset Levels.[29] Thereafter he was allowed only six boatloads of brushwood a year, to be taken out under view of the bailiff. Such rulings are a key to our understanding of the old wetland economy. Long before our modern preoccupations with ecology, the men of the wetlands harvested the wealth of their so-called wilderness with a sophisticated understanding of the need not to over-exploit its resources.

Because of the hostile elements facing farmers in the wetlands, mutual co-operation was essential, both in sharing the upkeep of flood-banks and drains and in administering a system of checks and balances to ensure that each person got a fair deal out of the common pasture. If overstocking took place, then everybody was the loser. In 1242 Geoffrey de Langelegh was summoned by the abbot of Glastonbury to explain why he now had 'one hundred and fifty goats and twenty oxen and cows beyond the number which he and his ancestors were wont always to have, to wit, sixteen oxen only'.[30] Annual 'drifts' were held, when cattle were rounded up, and excessive numbers were impounded and released only on payment of a fine. By Tudor times the Fen commons were subject to sophisticated management, controlled by the parish order-makers, who in turn appointed field reeves and fen reeves. These kept a close check on the taking in and pasturing of cattle by outsiders through a system of branding and regular drifts. During the reign of Edward VI, a code of fen law was drawn up, which remained fully operative in the Lincolnshire fens until the eighteenth century. Penalties were levied for putting diseased or unbranded cattle on to the fen, leaving animals unburied for more than three days, and allowing dogs to harass cattle on the moor. No reed was to be mown for thatch before it had two years' growth. No swans', cranes', or bitterns' eggs could be taken from the fen.[31] In the sixteenth century, Otmoor was similarly controlled by a moor court, upon which the local villages were represented by two 'moor men'.

The fixing of dates was critical in preventing over-exploitation. As early as

1534, a closed season for wildfowling, between May and August, was instigated in the Cambridgeshire fens. All inhabitants of the manor of Epworth on the Isle of Axholme had the right to set bush nets and catch white fish on Wednesdays and Fridays. Stocking on some fen commons took place no earlier than old May Day, to ensure against overgrazing. Lammas land was pasture open to commoners from Lammas, or Loaf Mass, 1 August. Lammas meadows still exist, as at Twyning near Bredon in Worcestershire and, more famous, North Meadow beside the Thames at Cricklade.

Such co-operative management existed even on some wetlands which were not commons. On the Derwent Ings in Yorkshire, the subdivision of the land into small hay plots and the subsequent pasturing were administered by a court leet, which annually appointed 'Ings Masters', who still manage the pastures at East Cottingwith and Newton. In the Wheldrake Ings account book for 1868–1934, it is specified that the meadows be mown on the dates appointed by the Ings masters, and that thereafter a carefully controlled number of cattle, branded with a W, may be pastured until the autumn, when they are taken off on 'Ings Breaking Day', a custom which is still observed.

The water, flooding over the pastures in winter and oozing up through the summer marshes, held the key to these balanced systems. The black waters of the fen halted the plough, thereby limiting the expansion of economic growth which a fen parish could sustain. Organized common grazing on wetland pastures blunted incentives to enlarge adjacent farms, and also prevented the selective breeding of livestock. But technologies of drainage, refined with each succeeding century and reaching a climax in our own day, removed the subtle checks and limitations of the old wetland systems. As the water ebbed away, so the spell was broken.

It is a mistake to be too naïve about the old wetland commons. They were open to abuse, and their system of controls did not always work. Overstocking took place on the Somerset Levels in the Middle Ages, aggravated by the rights of some commoners to take in cattle from outside the Levels for a fee. The Wallingfen court in the Vale of York set an upper limit for animals on the common in 1636 which was way above the actual carrying capacity of the land.[32] In the eighteenth century, Thomas Stone noted that West Fen in Lincolnshire was 'perfectly white with sheep'.[33] It was no doubt grossly overstocked. The damp conditions would exacerbate such a situation. Before the Brue valley drainage in 1770, 10,000 sheep rotted in one year in the Somerset parish of Mark.[34]

Nor were the commons some kind of pastoral socialist Utopia, open to all comers. By the Middle Ages, Otmoor was already a restricted common, jealously guarded by the inhabitants of its 'seven towns'. Albert Pell, a fen landlord, wrote in the mid-nineteenth century: 'The vulgar idea of the general public having rights of any kind on the waste or commonable land was never for a moment admitted.'[35]

Under Cromwell, the truly radical Diggers demanded that all commons and wastes should be cultivated by the poor in communal ownership. When they began to dig up waste land on St George's Hill in Surrey, they were driven off by local farmers, who almost certainly included small peasants angered at the usurpation of *their* common rights.

The agricultural improvements resulting from drainage did open up the possibility of betterment for the small man, as well as for the great landowners. In the Middle Ages a degree of drainage which allowed the conversion of pasture to meadow provided livestock farmers with that most precious of commodities, winter fodder. In 1606 the lord and his tenants co-operated to reclaim part of the moor at Cossington in the Somerset Levels. In the 1830s the farmers of Burwell fen began to realize that they were missing out on the prosperity achieved through drainage by their neighbours at Swaffham. Protagonists of seventeenth-century fen drainage pointed out that a fat ox was better than a well-grown eel, and a tame sheep more use than a wild duck. Underlying local issues concerning the draining of the marshy commons was the national issue of the need for food. Bad harvests between 1593 and 1597 were the prelude to the great fen drainage projects of the seventeenth century, at a time when England's growing population was increasingly concentrated in urban or rural industrial centres which were not self-sufficient. The farming achievements of the eighteenth and early nineteenth centuries were also motivated by the need for more food. By 1870 Europe was able to sustain a rapidly rising population, largely from its own resources: the agricultural revolution, including the critical part which drainage had played in it, had worked.

But back on the marshes and fens, who was really to profit from this continual process of ever more intensive cultivation? Thomas Fuller, writing in 1655, had the answer: 'Grant them drained, and so continuing; as now the great fishes therein prey on the less, so the wealthy men would devour the poorer sort of people . . . and rich men, to make room for themselves, would jostle the poor people out of their commons.'[36] For small farmers and commoners, drainage generally meant at best higher rents, at worst dispossession. Those undertaking the drainage were quick to stake their claim to the best bits of land. Thus James Bentham writes of the Fens: 'The smallest spots, however, scattered or remote, which first showed themselves above the surrounding waters, were eagerly seized upon by these watchful discoverers, and claimed as part of their allotted reward.'[37]

It was a similar story in Somerset in the same period, worsened by the fact that access to the moor along the old drove-ways was severed by the new drainage ditches. An often quoted rhyme seems to sum it up:

> They hang the man and flog the woman
> Who steals the goose from off the Common;
> But let the greater villain loose
> Who steals the Common from the goose.

Not that the goose ever had much of a deal. Live plucking was normal practice by the commoners, in order to ensure quills of the best quality.

## The People of the Wetlands

Taming the flood necessitated taming the people of the flood-lands. Outsiders, who generally initiated the drainage, were as unimpressed by the men of the marshes as they were by their stagnant swamps. Camden described fenmen in 1586 as 'rude, uncivil and envious to all others whom they call Upland Men; who stalking on high upon stilts apply their minds to grazing, fishing and fowling'.[38] Lieutenant Hammon, writing in 1635, went further: 'I think they be halfe fish, halfe flesh, for they drinke like fishes and sleep like hogges.' The people of Ely 'have but a turfy scent and fenny posture about them, which smell I did not relish at all with any content'.[39]

'Fenmen, disgusting representations of ignorance and indecency!' exclaimed the judge in the Littleport riots in 1816. In the same period, Arthur Young, subsidized by the big landlords to promote agricultural improvement, put his finger on what must have been a general attitude, when he described cattle-stealers in the Lincolnshire fens: 'So wild a country nurses up a race of people as wild as the fen, and thus the morals and eternal welfare of numbers are hazarded and ruined for want of an inclosure.'[40] They certainly were a rough lot. Thomas Stone in 1794 described Deeping fen as a frequent resort of cattle-thieves. Between the 1740s and the 1820s, Romney Marsh was openly terrorized by armed gangs of smugglers. Richard Gough, in his 'History of Myddle' in Shropshire, written in 1700, describes how resentment aroused by increased rents for peat cutting following drainage improvements bubbled over into violence. The agent of Sir Edward Kinaston approached a certain Clarke for rent, when he was 'cutting peates on Haremeare Mosse. . . . But one of Clarke's sons with a turfe spade, which they call a peate iron, (a very keen thing,) struck Sir Edward's man on the head and cloave out his brains. The bayliffe fled.'[41]

In the 1860s the first policeman ever sent to the fen village of Wicken was killed when he tried to break up a Saturday night brawl. His body was wheeled off in a peat barrow and cremated in the local brick kiln.

Ever since opposition to drainage in the seventeenth century, the men of the Cambridgeshire fens were known as 'fen tigers'. Their women folk must have been equally formidable. In 1632 'a crowd of women and men, armed with scythes and pitchforks, uttered threatening words' to anyone attempting to drive their cattle off Holme fen.[42] In 1539 Sir Richard Brereton decided to enclose and drain the Dogmore, a marshy common near Prees in Shropshire, which he had bought from the bishop of Lichfield. The bishop was harangued by 'fourtie wyfes of Prees', one of whom 'rudeley began to take his horse by the bridell whereat the horse sprang aside and put the Bysshop in danger of a fall'. Twelve years later, Brereton again went to the Dogmore, to appease 'great tumults of the

Tennants ther gathered together'. The local justice of the peace excused himself, saying he was 'dysseasid of styche'.[43] The prospect of an armed mob, including that monstrous regiment of 'wyfes', must have been enough to bring on an immediate headache. A riot in 1694 at Hatfield Chase in the Isle of Axholme is described by George Stovin, who was born the year after the events described: 'Whilst the corn was growing, several men, women and children of Belton and among others the said Popplewell's wife encouraged by him—in a riotous manner pulled down and burnt and laid waste the thorns and destroyed the corn.'[44]

The 'thorns' must refer to the new enclosure hedges planted on the commons. In Hatfield Chase today, you can see a sign put up by a local farmer with a sense of humour. It reads not 'Beware of the Dog', but 'Beware of the Woman'.

Such tough independent people must have posed a threat to both central and local government. Just as the Biesbosch on the Rhine delta was a centre for the Dutch underground opposition to Hitler, so the English wetlands have a long history as centres of resistance. Dio Cassius describes the difficulties with which the Romans subdued the ancient Britons, who hid in the marshes 'with their heads only out of the water!'[45] Alfred the Great led the resistance against the Danes from Athelney in the Somerset Levels; and although every schoolchild knows that William of Normandy conquered England in 1066, he did not succeed in subduing Ely and the surrounding fens until 1071, when Hereward the Wake submitted. Marshes have always been easy to defend. Romney Marsh was flooded as a defence against both Napoleon and Hitler, and Calais was lost in 1557 in part because the sluices were not opened in time to flood out the besiegers.

It would be a mistake to exaggerate the role of wetlands in national insurrections. Nevertheless, three marshland villages in Essex led the way in the Peasants' Revolt of 1381; Sedgemoor will always be associated with Monmouth's Rebellion; and agitations against fen drainage played a small, but significant, part in both the career of Oliver Cromwell and the origins of the English Revolution.

The place which seems above all to encapsulate the spirit of the Derwent Ings is Aughton church, standing alone in the marshlands, its churchyard lapped by floods each winter and haunted by the bubbling call of the curlew in spring. On the church wall is carved the watery symbol of a newt. Small boys in Yorkshire and Worcestershire, going out with their jam jars to collect newts and tiddlers, still talk of going out after 'asks'. The newt at Aughton is the emblem of Robert Aske, and it was from here that Aske set out in 1536 to lead the Pilgrimage of Grace against the religious reforms of Henry VIII. Aske's main aim was a return of the smaller monasteries, but his appeal included requests to halt enclosure and drainage. He typifies the marshman's feudal protest against central authority, together with his longing, not for a new order, but for a return of the old.

The wetlands are lost landscapes. Just as they defy access, they defy organization by outsiders. Even long-drained regions, such as Longdon Marsh in

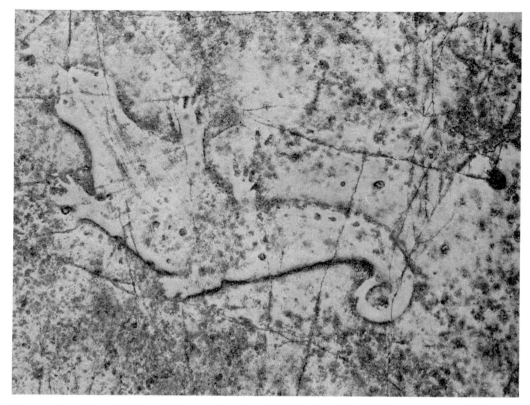

A newt, the emblem of Robert Aske, carved on the wall of Aughton church, Derwent Ings, Yorkshire.

Worcestershire, are easy to pick out on the Ordnance Survey as 'holes' on the map. Cul-de-sacs skirt warily down to them and then peter out. Some, such as the Somerset Levels, Chat Moss, and Hatfield and Thorne Wastes, are visible from motorways, from which there seems to be no exit whereby they can be reached. Approached more closely, they still challenge the intruder. Otmoor and Hatfield Chase are both encircled by moats of ditches, crossed in the case of Hatfield by only one bridge. If you do venture by car on to the edge of Otmoor, there is the feeling that you will be unable to turn around in the narrow space between the dykes, or may get stuck up to the axles in mud. The single rough road across Simonswood Moss near Liverpool is barred at either end by the intimidating iron gates of the Knowsley estate. Romney Marsh, which lacks a central inaccessible fastness, is crossed by a maze of switchback lanes which seem determined to throw off even the most diligent map-reader. In Somerset, the old drove-ways still branch off the main routes into the moors, like spines on a stickleback. These are truly the landscapes of the 'No Through Road'.

Straddling boundaries, some wetlands still defy comprehensive administration. Romney Marsh is shared by Kent and Sussex; what is left of the great moss system of the Mersey valley is carved up between Lancashire and Cheshire and the urban authorities of Manchester, Liverpool, and Warrington. Hatfield and Thorne Wastes are bewilderingly divided among Nottinghamshire, Yorkshire,

Lincolnshire, and South Humberside. The inhabitants would probably not admit allegiance to any of these, and huge signs proclaim the Isle of Axholme—known locally as 'The Isle'—as if it were an independent State. If such is the character of these places now, what must they have been like when both men and livestock could only get around them by boat, and parishes such as Dogdyke in Lincolnshire had in the eighteenth century 'not two houses communicable for whole winters round'.[46]

For all their imperfections, the old wetland commons had a certain self-sufficiency and self-containment which provide a standard against which to judge the enthusiastic, never-satisfied ambitions of the agriculturalists and drainage men who set out to exploit them. Traditional management of the marshes was tuned to the finest nuances of the local water-table. Each wetland evolved a landscape character as individual as the spirit of its people was independent. A few marshes in western France are still managed very strictly as commons. To visit them is to gain an insight into what many of our own wetlands must have been like. The Marais Communal of Curzon lies in the lap of low scrubby hills, like a green sea of stillness. It is quite without the trees or hedges which enclose all the country around it. Cattle, herded down to it along drove-ways used from time immemorial, slowly graze across its moist levels. From the steady centre of this tranquillity flickers the occasional silver of snipe or redshank, like fish rising from the still heart of a pool. The real beauty of such places is not their actual visual components, but the system which underlies them: the harmony between man and nature which they represent.

Andrew Motion's recent poem 'Inland' describes how a society, as much as an ecology, was overturned by drainage projects in the seventeenth-century fens. In this extract, a fen villager watches the arrival by boat of the men who are going to change his life:

> Sun flicked round the bay,
> binding the outline of farms
> to their reflections in grey
> bands of light. The marsh
> always survives. Always.
>
> Cattle stirred in their shed,
> uncoiling sweet whisps
> of breath over my head;
> fresh shadows spilt down
> their flanks and spread
>
> across water to flake
> into shrinking fragments
> over the strangers' wake.
> Their boat put down
> some men; one staked

its prow into our land,
waded towards us
over the grass, and
lifted one arm. Our world
dried on his hand.[47]

That world was one of many small men co-operating in order to survive. In the battle to save West Sedgemoor and the Derwent Ings in the 1970s and 1980s, the large number of small landowners was to militate against the efficiency with which large-scale drainage schemes could be organized. In 1794 Billingsley described the Somerset Levels as 'destitute of gentlemen's houses';[48] and the 1580 muster returns for Holland in Lincolnshire lamented 'the want of gentlemen here to inhabit'.[49] Charlton-on-Otmoor means 'town of the churls'. A 'churl' was a free peasant (note the slur implied by present dictionary usage), and Charlton never had a resident squire, being dismissed in eighteenth-century diocesan returns as having 'no family of note'.[50] The men of Charlton must have cherished their independence, especially when they looked at the fate of the neighbouring village of Noke, which, it was said, was lost by Lily, duchess of Marlborough, at cards.

Recent reports on current or just completed land-drainage schemes emphasize the trend whereby large farmers accrue the benefit much more commonly than small men. The theory behind such schemes is that ambitious large farmers will set an example, which will encourage their small backward neighbours. The latter are described in all current cost–benefit reports of the Ministry of Agriculture as 'laggards'. The assumptions behind this unfortunate word go back a long way. In 1652 Dugdale described fenmen as a 'lazy and beggarly people'. Billingsley castigated the farmers of the Somerset Levels in the late eighteenth century thus:

The possession of a cow or two, with a hog and a few geese, naturally exalts the peasant in his own conception, above his brethren in the same rank of society. . . . In sauntering after his cattle, he acquires a habit of indolence . . . and at length the sale of a half-fed cow or hog, furnishes the means of adding intemperance to idleness.[51]

No doubt such people must have seemed slow-witted. In order to counteract the effects of malaria, they were frequently doped with opium, which was sold over the counter in the village shop or grown in the fens (where Poppy Hill and Poppy Farm still exist as place-names). Where *Cannabis* was grown as an important fibre crop in the Cambridgeshire fens, the workers in the hemp fields were known to become exceedingly drowsy. Moreover, the isolation of the men of the wetlands led to inbreeding. In 1870 the geologist de Rance commented on the number of idiots in the Lancashire Fylde, which resulted from 'the dislike of the people to marry outside the district'.[52]

The laggards were constantly encouraged to improve for their own good. Arthur Young, as usual, has the last word. Here he describes the wetlands of Lincolnshire: 'Fens of water, mud, wildfowl, frogs and agues have been converted to rich pasture and arable worth from 20 shillings to 40 shillings an acre: health improved, *morals corrected* and the community enriched'.[53]

The history of drainage since the sixteenth century has seen the decline of enforced co-operation in sharing a resource, in the face of individual private enterprise. The big man has got bigger, and the small man smaller. This thought brings us bang up to date. The great agricultural revolution of our own times, in which drainage has played no small part, has accelerated the decline of the small farmer just as surely as it has imperilled the ecological system previously sustained by communal wetland management; and, at the last, it has begun to destroy the basic resources of the land, as ever deeper drainage has created mineral problems in the soil, wastage of peat, and an increasing dependence upon pumping. Man's attempts to tame the flood have not always progressed smoothly. There have been frequent and major set-backs; and it may be that we are now on the threshold of a new era, in which, for the first time, leaders in society will make a conscious decision to allow the flood-waters in some areas to rise again.

With all the themes previously outlined in mind, it is time to make a brief chronological survey of that process, which began in the mists of time, and has made inevitable the problems and conflicts described in the remaining chapters of this book. We are witnessing only the latest episode in that long history, in which geography itself has been remade, and the landscape, now more than ever, is transformed not so much by the efforts of individuals, as by public policy and the stroke of a pen.

# 3

## The Winning of the Waters

### A History of the Fight against Flooding until the Post-War Era

TODAY, land-drainage operations are administered from tower blocks, with the aid of computers in the office and sophisticated machinery on the river bank. It all seems a peculiarly modern phenomenon. In fact, the technocrats of land drainage are heirs to one of the oldest forms of organized local government. In 1252 the 'jurats' of Romney Marsh are recorded as having had the power to repair the sea-wall and to control the ditches 'from time out of mind'.[1] Similarly, organized bodies of men walked the marshes of the Thames, setting out in the early morning mist to assess the repairs required, which were paid for by a charge known as 'wallscot' or 'scottage'. Those who escaped these earliest of water-rates could be deemed to have got off 'scot free'. In the 'Domesday Book', our first named drainage specialist makes his appearance. He lived on the Somerset Levels, and he was called Girard Fossarius, Gerard of the Drain.

The people of the Middle Ages inherited sea-walls and drainage channels which had survived from the Roman occupation. The Romans had gained expertise in flood alleviation and in irrigation projects from the Greeks and the Etruscans before them. They industriously developed these skills throughout Europe, and were quick to export them to their colonies.[2] The emperor Hadrian is commemorated in Britain not only by the engineering achievement of his famous wall, but also by the Car dyke, a catch-water drain encircling the western edge of the Fens, linking the river Cam at Waterbeach with the Witham near Lincoln.[3] On the Medway estuary in Kent, raised banks built by the Romans to keep out the sea lasted substantially until the eighteenth century,[4] and the extent of Roman reclamation appears to have been formidable. The military and political decline of the Roman Empire coincided with a worsening of the climate, and both contributed to a rising of the swamps in the Dark Ages. This must have reinforced the terror of such places, as portrayed in *Beowulf* and the chronicles relating to St Guthlac. None the less, settlement persisted in some of the wetlands. The priest Rumen, or Romanus, who was chaplain to King Oswy's wife in the seventh century, probably gave his name to Romney Marsh, large areas of which he owned and may have farmed. Charters of 875 and 972 concerning Longdon Marsh in Worcestershire give a picture of clearing and enclosure among streams and marshes, including a duck pond 'on ducan seathe', traces of which still survive under the M50 motorway.[5] Piecemeal drainage and reclamation in the Cambridgeshire fens, carried out by Saxon farmers around Wisbech and Elm, have been traced through the detective work of archaeologists, using aerial photography and careful analysis of the evidence on the ground.[6]

## The Medieval Church

With the dawning of the Middle Ages, the driving force for reclamation of the marshes became the Church. Many monasteries had settled for safety on secluded islands in the swamps. By the sixth century, a colony of holy men had gathered at Glastonbury in the Somerset Levels. As the confidence and prosperity of the monasteries increased, so did the enthusiasm with which the monks began to drain and develop the wetlands around them. This was the pattern in wetlands all over the country until the Reformation. It is hard to underestimate the impact of the monasteries, especially in the twelfth and thirteenth centuries.

The monks of Furness reclaimed the coastal marshes of Walney, with embankments incorporating beach pebbles. Cockersands Abbey drained and hedged part of the Lancashire Fylde. The surviving network of ditches on the Monmouthshire Levels beside the Bristol Channel is essentially that dug by the Benedictines in the twelfth century. In the thirteenth century the bishop of Durham instigated extensive drainage works along the northern shores of the Humber. The monks of Meaux were active in the Hull valley, and those of Fountains on the Derwent Ings. In 1180 the canons of St Thomas drained the flood-lands at Eccleshall in Staffordshire. Battle Abbey actively reclaimed the Pevensey Levels in Sussex, and on Romney Marsh a lead was given by the priors of Christ Church, Canterbury, and by the archbishops themselves. Just as the church at Aughton stands as a touchstone of the spirit of the Derwent Ings, so the little church of St Thomas à Becket at Fairfield represents all the romance and loneliness of Romney Marsh. Prior to drainage work in the 1960s, Fairfield was regularly islanded by winter floods. Sheep graze up to its walls, mellow with yellow lichen, and the reedy dykes which surround it are famous for their marsh frogs, whose operatic baritone can be heard a mile away on May nights. The dedication of the church to St Thomas is no accident. Becket may well have been closely involved in building the great walls of packed clay which still enclose the local 'innings', or sheep pastures. They must have added considerably to the wealth of the See of Canterbury.[7] Near Fairfield you can still see the innings of St Thomas. On a grander scale, the dog-toothed vault of Crowland Abbey arches like the jaw-bone of a mighty whale above the Lincolnshire fens, a monument to the riches which the monks harvested from the marsh.

The dog-toothed vault of Crowland Abbey arches like the jaw-bone of a mighty whale above the Fens, which the Abbey's founder St Guthlac believed to be the haunt of demons.

In a few cases the wetlands proved too much for them. The abbey of Otley on Otmoor was abandoned after three years, in 1141, as 'fitter for an ark than a monastery'.[8] But in general, the abbeys and their abbots fattened up together. Tithes of reed were reserved for the local priest on the Somerset Levels, and Chaucer's monk cast an entirely practical eye on the local birdlife: 'he liked a swan best, and roasted whole.' The holy men, whose heirs, such as the Carmelites on the Derwent Ings, now venerate God's wilderness which washes up to their walls, began to compete with each other as to who could plunder it the most. In 1305 the abbot of Thorney in the Fens complained that the abbot of Peterborough 'lately by night raised a dyke across the high road', and so cut off the former's access to corn and pasture.[9] On the Somerset Levels, intermittent war was waged throughout the Middle Ages between successive abbots of Glastonbury and bishops of Wells. They were forever breaking up each other's fish-weirs and quarrelling over competing interests in pasture and peat cutting. In 1278 the abbot's men destroyed a piggery belonging to the bishop in Godney Moor, and

The church of St Thomas à Becket at Fairfield epitomizes all the romance and loneliness of Romney Marsh. Photo: Jo Nelson.

again in 1315. In 1326 somebody set fire to the peat moor in the Brue valley, with the idea of burning Glastonbury Abbey. The bishop followed up this preliminary scorching with a promise of eternal fire for the abbot of Glastonbury, upon whom he pronounced sentence of excommunication for the sin of damaging his property.[10]

## The Courts of Sewers

Such piecemeal, not to mention conflicting, management of the marshes was no way to organize and control the ever-threatening flood-waters; and from the mid-thirteenth century, the responsibility for land drainage and reclamation from the sea began to devolve upon successive 'commissions of sewers', which were answerable to central government. A 'sewer' was a straight cut, the kind of geometrical channel beloved by modern engineers, and did not have the connotation of foul water which it has today. The first commission of sewers was set up in Lincolnshire by Henry de Bathe in 1258. Like fenland engineers 400 years later, de Bathe went for advice on procedure and administration to the heartland of organized land drainage, Romney Marsh. The subsequent courts of sewers were steadily reinforced by successive legislation, culminating in 1532 in Henry VIII's Statute of Sewers, just at the moment when the power of the monastic lords of the levels was broken by the Reformation. These courts were to survive, incredibly, until 1930. With the growth of commissions of sewers in the Middle Ages, the role of the professional layman in such matters began to increase. In 1390 a commission was appointed to inspect and repair flood-banks and dykes on the Thames marshes between Greenwich and Woolwich. It included among its number the king's clerk of works, no less than Geoffrey Chaucer.[11] The last major drainage operation in the medieval period, however, was instigated by a churchman. John Morton, later to become Henry VII's lord chancellor, familiar to every schoolchild for his tax levies of 'Morton's Fork', organized the construction of the channel still known as Morton's Leam, when he was bishop of Ely between 1478 and 1486. This ambitious piece of engineering, extending for 12 miles between Peterborough and Wisbech, survives today, although the tower Morton built from which to watch over his work-force crumbled away in the early nineteenth century.

By the end of the Middle Ages, the marshes in Kent and Sussex had been sufficiently reclaimed to reveal an abiding characteristic of such operations: that in certain circumstances land drainage contains the seeds of its own destruction. In the mid-fifteenth century, floods and silting doomed the old port of Pevensey; and further along the coast, on Romney Marsh, a series of catastrophic storms culminating in 1287 obliterated the towns of Old Winchelsea and Broomhill. Although climatic deterioration in the later Middle Ages certainly worsened the situation, these disasters were not simply the haphazard expression of hostile Nature. They were made inevitable by the meddling of man. At Pevensey,

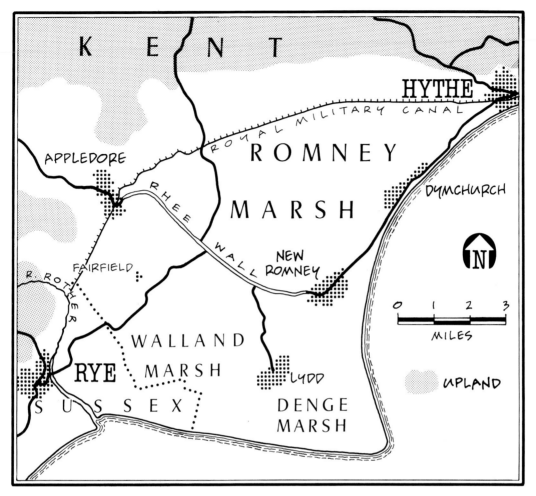

Romney Marsh.

reclamation of the adjacent estuary had reduced tidal scouring, which had previously kept the river mouth open. In consequence, the water, unable to escape through the blocked outfall, flooded the land, and navigation up the river was also prevented. Successive new channels cut in 1402 and 1455 failed to remedy the problem, and Pevensey Castle, which still rises dramatically above the marshes whose creation ensured its demise, was abandoned.[12] On Romney Marsh, silting of river mouths was worsened by the problem of peat shrinkage. By the twelfth century, increasingly elaborate drainage schemes had led to contraction of the peat, thereby causing the land to drop ever lower in relation to the menacing waters of the English Channel. When the banks finally broke, the sea captured both arms of the river Rother, and created the present estuary south of Rye. This cat-and-mouse game between engineers and the elements was to become a major theme in the next great age of wetland reclamation, which began under the Tudors and reached its climax in the middle of the seventeenth century.

Following the dissolution of the monasteries in the 1530s, the flood-waters began to rise again, since there were no monks to man the sluices and dig the drains. At least, this was the opinion of Dugdale, writing a hundred years later; but it should be remembered that he was one of a long line of propagandists for land drainage, and that, while localized deterioration must have taken place, there is also ample evidence of the activities of the courts of sewers and of individual enterprise by secular landlords. In 1539 the gentry around Newhaven diverted the Sussex Ouse to improve the drainage of the estuarine marshland and to capture navigation and trade from their neighbours at Seaford. During the reign of Elizabeth, the Wealdmoors in Shropshire were a battleground between rival landlords intent on drainage and enclosure. In 1576 Thomas Cherrington complained in the Queen's Council of the Marches, that Thurston Woodcock, lord of Meason, had assembled a gang armed with long staffs and billhooks, and had forcibly ploughed and then enclosed a piece of his waste ground with a ditch. Clearly Cherrington was not above such tactics either, for he seems to have destroyed the ditch. In 1583 the Woodcocks were back 'with divers . . . desperate and lewde persons' who 'in riotous manner dug . . . one myghty diche more like in truthe a defence to have kepte owte some forren enemyes than an inclosure to keepe in cattell'.[13]

## The Battle for the Fens

Towards the end of Elizabeth's reign, covetous eyes began to be cast on far grander prizes. The new card in the pack was foreign technology. In the reign of Henry VIII, Italian engineers had recovered Combe Marsh near Greenwich; and under Elizabeth, Plumstead and Erith marshes by the Thames were drained with the help of Dutch engineers and workmen.[14] In 1575 a certain Peter Morris of Dutch extraction obtained a licence from the queen to employ engines, or mills, for draining; and in 1589 Humphrey Bradley, who, despite his name, came from Bergen op Zoom in the Netherlands, presented a treatise to Elizabeth's chief minister, Lord Burghley. It proposed nothing less than the reclamation for her kingdom of an area 70 miles long and 30 broad, equivalent to a whole new county, the 'Great Level' of the Fens. Shrewdly, in view of the trouble such ambitions were to cause her successors, the queen turned Bradley down. He went off to France, where, as Master of the Dykes for Henry IV, he supervised the draining of the great Poitevin marshes north of La Rochelle. But before she died, Elizabeth I signed an Act of Parliament in 1600 'for the recovering of many hundred thousand Acres of marshes'. The battle for the Fens was on.

The Stuart kings were forever short of money. The career of James I was marked by ingenious methods of raising cash, by fair means or foul; and similar expropriation was to lead his heir, Charles I, to his downfall. In such circumstances, the new engineering which could apparently transform wetland wastes into sources of valuable crops to be sold to a growing population, must

'Covetous and bloodie Popham', an early drainer of the Fens, lies in state in Wellington church, Somerset.

have seemed like something for nothing, and therefore irresistible. Francis Bacon advised King James to hold on to his royal wastes and hunting forests for exactly this potential; and, as if to confirm the good sense of such drainage enterprises, a series of bad winters between 1607 and 1613 created some of the worst floods in living memory. In Somerset 'the tops of trees and houses only appeared . . . as if, at the beginning of the world, townes had been built in the bottom of the sea'.[15] In the Fens, mothers abandoned their children 'swimming in their beds, till good people, adventuring their lives, went up to the breast in the waters to fetch them out at the windows'.[16]

For the flood to yield up its riches, two things were required: a competent engineer and plenty of capital. To obtain the latter, there emerged a peculiarly modern group of businessmen who called themselves 'undertakers' or 'adventurers'. An undertaker was one who contracted to 'undertake' a drainage scheme; an 'adventurer' was one who 'adventured' his capital on such an undertaking. The security of both was the promise of a large proportion of the land after the drainage operation had been successfully completed.

In 1605 Lord Chief Justice Popham, prosecutor of Guy Fawkes, Ralegh, and the Queen of Scots, 'undertook' to drain the fen at Upwell. He has left behind him a flamboyant monument in Wellington church, Somerset, and the channel known as Popham's Eau in Cambridgeshire, which was abandoned at his death in 1607. The judge's real memorial, however, is his reputation. In 1606 James I received an anonymous letter accusing 'covetous and bloodie Popham' of ruining the poor people of the Fens.[17] The commoners, who had everything to lose from undertakings such as his, were firing an opening shot. The wetlands and wastes of England were soon to be loud with their tumults.

In 1618 James made his first move in Somerset. He decided to drain King's Sedgemoor, which the Crown had inherited entire at the dissolution of Glastonbury Abbey. Three years later, in 1621, the king declared that he would himself undertake the drainage of the Fens for a recompense of 120,000 acres; and in that year, there arrived in England a Dutchman who was destined to become one of the greatest architects of the English landscape. His name was Cornelius Vermuyden.

In 1625 the old king died, having achieved no effective drainage operations. Vermuyden had been occupied with rebuilding the Thames flood defences at Dagenham, which, according to an inquiry of 1623, he left 'in a worse condition than it was before'.[18] From this bad beginning, Vermuyden turned his attention to the great wetland system around Hatfield Chase, south of the Humber. It was here that he made his name, carrying out the first really ambitious operation of its kind in the country, which was to be a blueprint for his 'Great Design' for the Fens. It is also at Hatfield that we first get a clear glimpse of the tricky controversial personality of Cornelius Vermuyden.

The name 'Trent' comes from the Celtic word 'Trisanton' meaning 'trespasser', which describes the wandering nature of the river. The river Trent has always flooded, and nowhere more so than on the levels between where its own waters and those of the Yorkshire Aire flow out into the Humber. Apart from the little Isle of Axholme, this basin must all have been a flooded fen roughly 8 miles wide by 12 miles long. It remains one of the least known, and, despite much intensively farmed land, one of the wildest regions of lowland England. At its heart lie two raised mires of deep peat, wildernesses of scattered birch, where adders sun themselves among the fern, and amber dragonflies haunt the peaty pools. The

Adders still sun themselves on Hatfield Chase.

larger mire, Thorne Waste, approached by surmounting the incongruous dere-
liction of Moor Ends Colliery, stretches out as far as the eye can see, an
astonishing 6,000 acres of untamed wetland. To its south lies the smaller Hatfield
Moor, still quite large enough to get lost in. Ruined cottages nestle among its
birches, and miners, with guns over their shoulders, roam its maze of tracks on
the look-out for duck and rabbit. At dusk the air is filled with the eerie chuckle
of nightjars. In its very centre, accessible only by an earth track, is the ancient
manor of Lindholme, which was, from the Middle Ages, a royal hunting-lodge.
It was from here that the Stuarts sized up the potential of the area for exploitation.
In 1626 Charles I signed an agreement with Vermuyden for the drainage of
Hatfield Chase, for which the latter would be awarded one-third of the land
drained. Work started immediately, and within three years the scheme was
completed. The dykes he dug to create the farmland around Hatfield Moors can
still be seen, harbouring the aquatic flora of the ancient fen: butter-yellow
bladderwort and the feathery spires of mare's-tail.

In January 1629 the king knighted Vermuyden at Whitehall, and a month later
he sold him his royal manor of Hatfield for £10,000 cash down and an annual
interest of £195. 3s. 5d. and a red rose. But matters had not gone as smoothly as
all that. Even the knighthood was not the honour it might seem, for James I
had instigated the practice of charging for knighthoods, and Charles I had
compounded his profit by fining those who had the temerity to refuse. Furthermore, since the Crown had earlier inherited Hatfield subject to the maintenance
of rights of common, it was not entirely the king's to either sell or drain. This
was something which the inhabitants were not going to overlook as lightly as
Charles had done. A lawsuit of April 1629 between Vermuyden and the commoners attests how the people of Torksey

came unto the workmen and beat and terrified them, threatening to kill them, if they
would not leave their work, threw some of them in the river and kept them under water
with long poles, and at several other times, upon the Knelling of a Bell, came to the
said works in riotous and warlike manner, divided themselves into companies, to take
the workmen and filled up the ditches and drains, made to carry away the water, burned
up the working tools and other materials of the Relator and his workmen, and set up
poles in the form of gallows, to terrifie the workmen and threatened to break their arms
and legs, and beat and hurt many of them and made others flee away, whom they
pursued to a town with such terror and threats, that they were forced to guard the
town.[19]

Reports in the previous year that a local man had been killed by the Dutch work-
force make it clear that the battles over the digging of the ditches were far from
one-sided. Worse was to follow. The inhabitants made it clear that their commons
had been reduced to between half and a third of their former size. While
propagandists of drainage, such as Dugdale, admired the corn and the oil-seed
rape which could now be sown on the drained land, they failed to appreciate
that the people of Axholme already grew sufficient corn for their needs on the

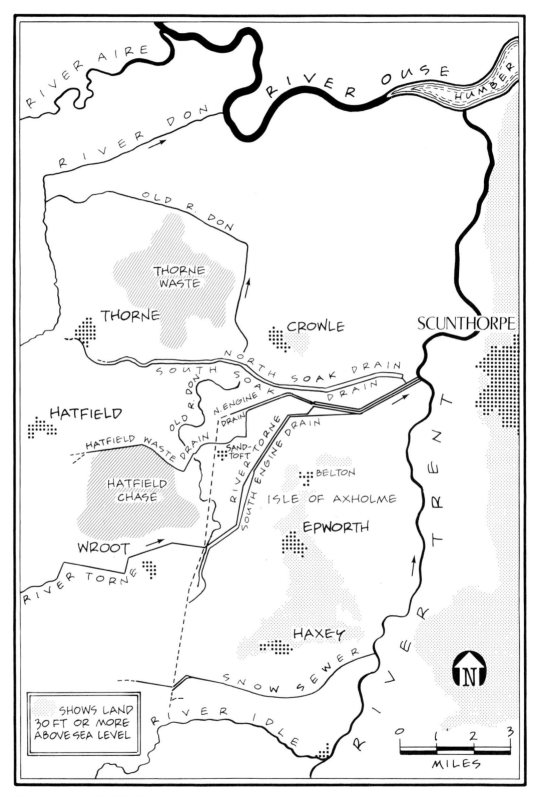

The drainage of Hatfield Chase.

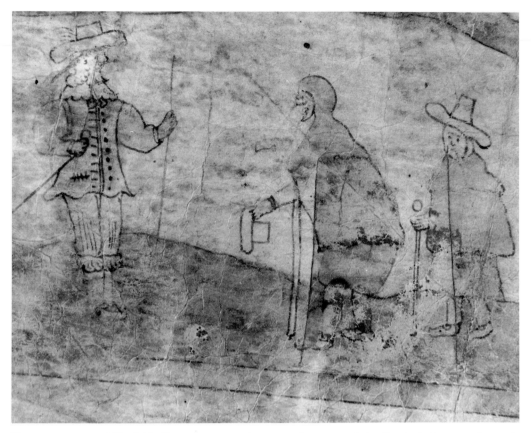

Two commoners of Hatfield Chase greet a gentleman, perhaps one of the Dutch drainage engineers. In reality they were rather less deferential. From the map made in 1639 by Thomas Arlebout, mariner, to mark the enclosure awards following the drainage of Hatfield Chase. Copyright: Nottingham University.

higher land. What the people wanted on the fen was what they had already: grazing. The ignorance with which the outsiders set about overturning a perfectly satisfactory agricultural economy at Axholme suggests parallels with the notorious ground-nut scheme instigated in East Africa in the late 1940s. In addition, the villagers of the north-west of the region complained, with justification, that the engineering works had simply sent the water down to flood *them* out. After much legal deliberation, the lord president of the Council of the North, the earl of Strafford, pronounced that Vermuyden must bear the cost of a major new channel, still called 'Dutch River', to rectify the situation. The project ended in financial disaster, and Vermuyden was temporarily imprisoned for not paying his debts. Catastrophic floods inundated the region in 1636 and again in 1697, exacerbated no doubt by deliberate sabotage by the commoners, but also caused by the insufficient capacity of the new channels and by peat shrinkage. The people remained as uncontrollable as the waters, burning down the Dutch settlement at Sandtoft during the Civil War and again in 1688.[20] Peace did not really reign again at Hatfield Chase until well into the eighteenth century.

However, long before these disastrous developments, Vermuyden was wiping the mud of Hatfield off his boots and casting around for greener pastures. From 1629 to 1632 he held leases in Malvern Chase, which included Longdon Marsh, and in 1632 he bought part of Charles I's share of King's Sedgemoor in Somerset. But Vermuyden never achieved effective drainage in the West. In 1636 he was accused by the king's agent for Somerset of fraud and duplicity, charges which were revived when he made another attempt to tackle the Somerset Levels in 1655. The Somerset commoners succeeded in fighting off most attempts at drainage where their contemporaries in the eastern counties had failed. After 1638 nearly two-thirds of the Somerset Levels were still unreclaimed, and even as late as 1769, the local drainage agent, Richard Locke, was stoned, and his effigy was burned 'by the owners of geese'.[21] Old habits die hard in the wetlands. In 1983 the descendants of these owners of geese were to burn the local conservationists in effigy.

The finest prize for the reclaimers remained the Great Level. In 1630 Francis Russell, earl of Bedford, agreed to undertake the drainage of the fens in Cambridgeshire, Huntingdonshire, and parts of Norfolk and Lincolnshire. In 1631 thirteen other business adventurers joined the earl, forming the Bedford Level Corporation and employing the services of Cornelius Vermuyden. Thus was inaugurated a period of extensive engineering works, culminating in the construction of the Old Bedford river in 1637. The adventurers then proceeded to bid for their profit. There was a general outcry. Many complained that the flooding was as bad as ever; others that Bedford had cheated them of their land. In 1638 Charles I intervened, and, re-engaging Vermuyden, declared himself undertaker. The king's ambitions were characteristically grandiose. Not only did he require a grant of 57,000 acres of the drained land, but, according to Dugdale, he intended to transform the village of Manea into a town to be called Charlemont, which would command the new river system. One can imagine the cloud-capped towers of Inigo Jones's elegant Baroque soaring above the Fens. As it is, Manea (pronounced Mainy) remains a tiny hamlet in the Ouse washes, haunted by the ghost of what might have been.

Events were moving fast to overtake all such enterprises. From his first arrival in the Fens, Vermuyden had been faced with the now familiar rioting. A drinking song called 'Powtes Complaint'—'powte' being a lamprey—circulated in the taverns:

> Come, Brethren of the water, and let us all assemble,
> To treat upon this matter, which makes us quake and tremble;
> For we shall rue it, if't be true, that Fens be undertaken,
> And where we feed in Fen and Reed, they'll feed both Beef and Bacon.
>
> They'll sow both beans and oats, where never man yet thought it,
> Where men did row in boats, ere undertakers bought it:
> But, Ceres, thou, behold us now, let wild oats be their venture,
> Oh let the frogs and miry bogs destroy where they do enter.

The beauty of many rivers lies in their long history of management by man. *Top*. Water crowfoot can be an indicator of ancient fords. *Bottom*. The mill-race is one of man's contributions to the landscape quality of the river.

In the early eighteenth century windmills were adopted throughout the Fens for pumping water.

The main drains on the Southern Fenland, including approximate dates of construction.

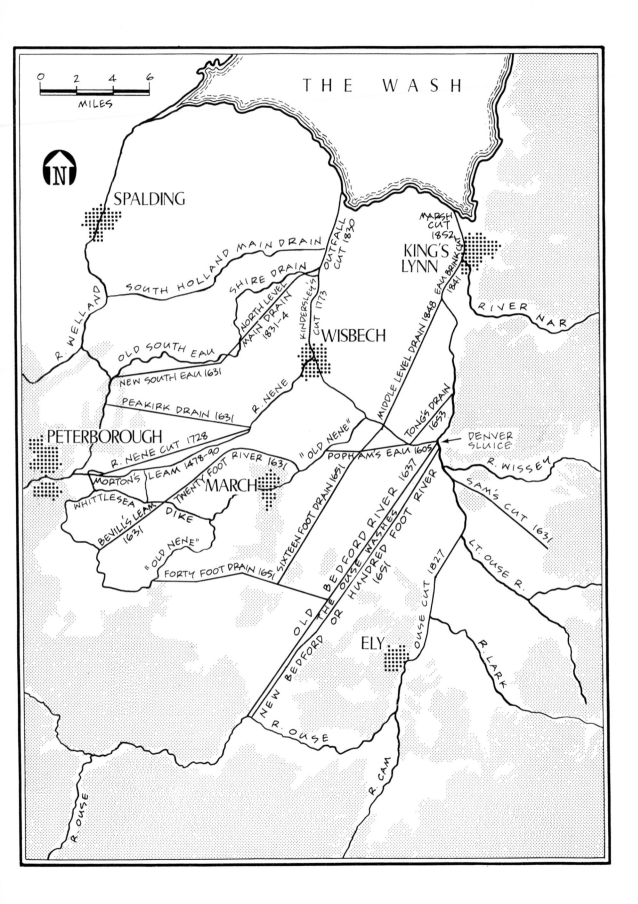

THE WASH

SPALDING

KING'S
LYNN

MARSH
CUT
1852

EAU BRINK CUT
1841

RIVER NAR

SOUTH HOLLAND MAIN DRAIN

SHIRE DRAIN

NORTH LEVEL
MAIN DRAIN
1831-4

OUTFALL
CUT 1830

KINDERSLEY'S
CUT 1773

WISBECH

R. WELLAND

OLD SOUTH EAU

NEW SOUTH EAU 1631

R. NENE

MIDDLE LEVEL DRAIN 1848

TONG'S DRAIN
1653

PEAKIRK DRAIN 1631

DENVER
SLUICE

PETERBOROUGH

R. NENE CUT 1728

POPHAM'S EAU 1605

R. WISSEY

MORTON'S LEAM 1478-90

TWENTY FOOT RIVER 1631

"OLD NENE"

SIXTEEN FOOT DRAIN 1651

OLD BEDFORD RIVER 1637

SAM'S CUT 1631

MARCH

WHITTLESEA

BEVILL'S LEAM
1631

DIKE

"OLD NENE"

FORTY FOOT DRAIN 1651

THE OUSE WASHES

NEW BEDFORD OR HUNDRED FOOT RIVER 1651

OUSE CUT 1827

LT. OUSE R.

ELY

R. LARK

R. OUSE

R. OUSE

R. CAM

N

0  2  4  6
MILES

A Dutch engineer, believed to be Cornelius Vermuyden.
Copyright: London Borough of Barking and Dagenham.

Charles I. National Portrait Gallery.

Oliver Cromwell, painted in the year of the execution of
Charles I. National Portrait Gallery.

The contemporary chronicler and advocate of seventeenth-
century drainage, Sir William Dugdale.

Away with boats and rudder, farewell both boots and skatches,
No need of one nor th'other, men now make better matches;
Stilt-makers all and tanners, shall complain of this disaster,
For they will make each muddy lake for Essex calves a pasture.

The feather'd fowls have wings, to fly to other nations;
But we have no such things, to help our transportations;
We must give place (oh grievous case) to horned beasts and cattle,
Except that we can all agree to drive them out by battle.[22]

Battle was what they settled for. The men of the Lincolnshire fens 'fell upon the Adventurers, broke the sluices, laid waste their lands, threw down their fences . . . and forcibly retained possession of the land.'[23] The northern fens were to remain the preserve of fishers and fowlers for another 150 years. In Cambridgeshire, Sir Miles Sandys, an adventurer whose capital had been sorely overstretched by the drainage projects, wrote to his son that if 'order not be taken, it will turn out to be a general rebellion in all the Fen towns'.[24]

General rebellion, indeed, but not only in the Fens. In 1638 the commoners found a champion of their causes in a local farmer whose career was ultimately to lead him far beyond the battles of the wetlands. In that year it was 'commonly reported by the commoners . . . that Mr Cromwell of Ely had undertaken, they paying him a groat for every cow they had upon the common, to hold the drainers in suit for five years'.[25] They appointed Cromwell their advocate at the commission of sewers in Huntingdon, and he ensured that a clause concerning the commandeering of common land was included in the catalogue of complaints known as the Grand Remonstrance presented to the king in 1641. The following year Civil War was declared, and drainage works fell into abeyance. In 1649 the war was over and the king executed, but that summer a surprising turn of events took place in the Fens. In May, Oliver Cromwell, erstwhile champion of the commoners, was named as one of the commissioners, together with the earl of Bedford, under a new Act for the Draining of the Great Level. He was to send a major of his own regiment to suppress the commoners' riots; and in 1654 he issued an ordinance to protect Bedford and his works, himself receiving 200 acres of the drained land as a reward. After intensive wrangling over terms and money, during which the adventurers declared that it was 'not fitt to depend upon Sir Cornelius Vermuyden any longer', the latter was re-engaged in 1650.[26] The following year, despite the midnight activities of 'the meaner sort of Burwell'[27] and other villages, he completed the New Bedford, or Hundred Foot, river, parallel to the Old Bedford built fourteen years before. The two massive channels run straight towards the Wash, enclosing a flood-land which still fills up in winter, as Vermuyden intended. They are his greatest monument. In 1653, after the remaining works had been completed by Dutch prisoners of war under Vermuyden's direction, a service of thanksgiving was held in Ely Cathedral; and

Leading figures in the battle to drain the Fens.

in the same year Vermuyden was employed by his earlier adversary, the lord protector, on diplomatic missions to the Netherlands. The fenmen and their fens were under control at last. Samuel Fortrey celebrated the achievement in verse:

> I sing floods muzzled and the Ocean tam'd
> Luxurious rivers govern'd and reclam'd.
>
> .     .     .     .     .     .     .
>
> Streams curb'd with Dammes like Bridles, taught to obey,
> And run as straight as if they saw their way.
>
> .     .     .     .     .     .     .
>
> New hands shall learn to work, forget to steal
> New legs shall go to church, new knees shall kneel.[28]

In the great struggle to extend cultivation to feed the growing cities, the peasantry had been defeated no less decisively than the Crown. From this turning-point, enclosure and drainage were to shape both lives and landscape in the English countryside, right up to the present day.

In 1658 Cromwell died, by appropriate irony, hastened on his way by malaria, probably contracted during his campaigns in the bogs of Ireland. About this time, Vermuyden vanished from public life, although it was not until 1677 that he died, full of years and riches, and was buried in St Margaret's, Westminster.[29] It is interesting to speculate on the possible reasons for his total obscurity during the intervening years of the Restoration. He certainly made many enemies, and not only among the commoners. In 1633 legal action was taken against him by his countrymen and co-adventurers at Hatfield Chase. His earliest ally had been Sir Robert Heath, who had promoted Vermuyden's advancement at court, and had even managed to get him out of prison. When Heath, a staunch Royalist, fled to France under the Commonwealth, Vermuyden appears to have expropriated Heath's share of a mine at Wirksworth, and Heath's son was still petitioning for his rightful property in 1652. With the new regime in 1660, there must have been those who had old scores to settle. They may also have been quick to point out that Vermuyden's 'Great Design' was already turning sour. He had been lampooned by contemporary dramatists. In Thomas Randolph's play *The Muse's Looking Glass*, a conversation takes place between an engineer named Banausus and a gentleman called Colax:

> BANAUSUS. I have a rare device to set Dutch windmills
>      Upon New-market Heath, and Salisbury Plaine,
>      To draine the Fens.
> COLAX. The Fens Sir are not there.
> BANAUSUS. But who knowes but they may be?[30]

How true this still rings for modern consultants on the look-out for schemes for which no justification exists. Vermuyden had faced remarkable difficulties, not least the age-old problem of clients who want the profit at the end of the day,

Cornelius Vermuyden's greatest monument, the two Bedford rivers, enclosing the Ouse Washes. The section below shows how the washland takes the winter flood-water. Photo: Cambridge University Collection.

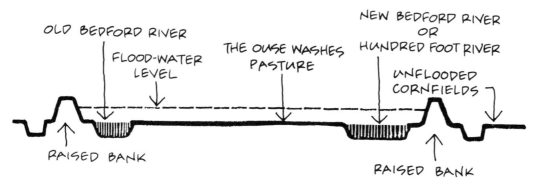

but who are not prepared to lay out sufficient capital to achieve it. Consequently, a linchpin of Vermuyden's scheme, a catch-dyke skirting the eastern edge of the fens, was abandoned, to be constructed only in the 1960s.

Such inadequacies were made far worse by something which Vermuyden could not have foreseen: peat shrinkage. The people might be made to kneel, but the elements were not quite so easy to muzzle. The very efficiency of any improvement speeded up the lowering of the land as the peat dried and 'wasted' through the activities of bacteria and fungi. Today the surface of the peat fens lies only a few feet above or below sea-level. Children playing in fen churchyards in the nine-teenth century were able to reach down and touch the coffins exposed by the wasted peat.[31] As the peat shrank, the critical outfall of the river Ouse into the North Sea inevitably began to silt up. By 1663 real problems were already apparent, and in 1673, four years before Vermuyden died, mighty floods inun-dated the Fens, forcing the inhabitants 'to save themselves in boats'.[32] By 1700 the full extent of the disaster had become clear, and in 1713, the Denver sluice, the key to the whole system, was washed out to sea. Not only in the Fens were the waters fighting back. An ambitious reclamation scheme on the south coast also came to nothing around this time. Between 1630 and 1646 Sir George Horsey had attempted to dam and drain the long tongue of water which still lies between Chesil Beach and the Dorset mainland. Winter storms swiftly obliterated his expensive engineering structures.[33]

### New Technologies

Just when it looked as if the flood was winning again, technology came to the rescue. In 1710 complaints were made in Norfolk about the cost of the new 'whirligigs';[34] and soon, travellers to the eastern counties were noting the number of windmills dotting the landscape. They were to become the crucial factor which saved the Fens from total inundation, and although not generally adopted in Somerset, they fast became a typical feature of such wetland landscapes as the Lancashire Fylde and the Hull valley. The story of the wetlands in the eighteenth century is one of gradual development of new technologies, cul-minating at the end of the century in the discovery of steam as an even more efficient pumping force than windmills. In 1692 Sir Thomas Fleetwood set out to drain Martin Mere in Lancashire, employing 2,000 men.[35] In 1755 storms destroyed the floodgates he had built, and renewed efforts to conquer the mere in the 1780s were crowned with permanent success only as a result of the use of steam-power in the following century.

Meanwhile, a less visible, but no less powerful, strategy for combating the waters was being devised. Every farmer knows that engineers may lower the levels of rivers for all they are worth, but that, without a follow-up operation of underdrainage in each saturated field, the real rewards for agriculture will never be harvested.[36] Open ditches and such ancient techniques as ridge and furrow are of limited effect.

Upstream from Leamington Spa, the river Leam flows lazily among water-lilies and tall bulrushes. Scraps of sedge and meadow rue still cling to its margins, the last remnants of a marsh which must once have inundated the whole valley floor. It was here, in 1764, that a Warwickshire gentleman made a discovery of the greatest significance. Mr Joseph Elkington of Princethorpe was faced with a problem. His sheep were suffering from foot-rot, and however many ditches he dug, he could not get the water off his fields. He was pondering his dilemma when a servant stopped by with an iron bar for making sheep hurdles. Mr Elkington rammed it into the bottom of one of his ineffective ditches and, to his astonishment, water burst up like a geyser. He had discovered a method of intercepting springs, and, using stone to seal his drains, he and others like him set about spreading the gospel of effective underdrainage.[37] Very soon farmers were using clay tiles, which, stamped with the word 'drain', were exempted in 1826 from the tax on other clay products. The clay tile and its descendant, the plastic pipe, were to take their place alongside the plough and the axe as among the major agents in the settlement of England.

## Eighteenth-Century 'Improvements'

In 1795 Parliament voted that King George III award Elkington £1,000 to carry out a survey of his achievements. It was in the reign of 'Farmer George' that drainage became more than ever in vogue, 'improvement' being all the rage. In the eighteenth century this word had two meanings, basically different faces of the same coin. 'Capability Brown', and later Repton were employed to 'improve' the beauties of the grounds around country houses. It seems curious that for all those hard-riding, hard-drinking squires, the eternal search for a status symbol should have taken the form of building temples to nymphs and dryads. None the less, the universality of this practice is attested by one wit who told Brown that he would like to die before him, so that he could have a look at heaven before Brown 'improved' it. The other meaning of the word, still current in farming circles, is to intensify agricultural production. This form of improvement no doubt helped to pay the fees of Brown and Repton, along with all the other bills. Invoking a doctrine of the 'spirit of place', they felled ancient woodlands and drained the marshes. Villages were razed to the ground, so that they did not disrupt the view from drawing-room windows, with the same enthusiasm with which the peasantry on the more distant corners of the estate were dispossessed of their wetland commons in the pursuit of productive farming.

The environmental contraditions implicit in all this activity scarcely occurred to anyone, of course. Sir Joseph Banks, the greatest naturalist of the age, founder of Kew Gardens and botanist-companion to Captain Cook, first developed his boyhood passion for natural history in East, West, and Wildmoor fens, which washed up to the foot of the Lincolnshire wolds, and so to the very gates of Revesby Abbey, the Banks's family home. In his old age, Banks presided over the destruction of these fens, supporting the drainage projects of John Rennie, according to *The Farmer's Magazine* of February 1807, against 'a party of uninformed people, headed by a little parson and a magistrate'.[38] His portrait hangs in the place of honour in the Boston office of the Anglian Water Authority.

Another botanist, William Roscoe, founder of Liverpool botanic gardens and commemorated by the genus *Roscoea* beloved of alpine gardeners, actually bankrupted himself as a result of a drainage scheme.[39] In 1793 Roscoe began work on Trafford Moss, part of the mighty Chat Moss, 2,500 acres of sphagnum, sundew, and bog asphodel. Roscoe's ambition was to drain the whole wetland, and to this end he organized ditching, marling, and importation from nearby Manchester of boatload upon boatload of human ordure, which was forked by hand on to the moss. One of Roscoe's ideas was a windmill plough, whose sails would actually churn up the bog. Unsurprisingly, in view of such projects, he was financially ruined, and his interest in Chat Moss was bought out by 1821.

No one was worse at making connections about the consequences of his actions than William Madocks who reclaimed the coastal marshes of the Traeth Mawr in North Wales.[40] His embankment across the Glaslyn estuary was completed in 1811 amidst much rejoicing and ox-roasting, only to collapse the following year. After its final reconstruction, it was to bear the main road and railway out of Portmadoc, named in honour of its founder, who, with sublime inconsistency, passionately espoused the fashionable ideals of picturesque landscape. The man who rammed a causeway across the front of the finest prospect of Snowdonia was actually given to carving breathless verses to the water sprites on the river cliffs at Dolgellau. As he imposed his geometrical grid of drainage ditches across the newly filled-in estuary of the Traeth, it occurred to Madocks for a brief, but anxious, moment that the whole project resembled 'Dutch gardening'; but in no time the poet Shelley arrived to help him with his endeavours, declaiming on the 'poetry of engineering'. Only one man could see the situation clearly: Thomas Love Peacock, who described the scenic effect of Madocks's project in his novel *Headlong Hall*: 'The mountain frame remains unchanged, unchangeable: but the liquid mirror it enclosed is gone.'[41]

## Enclosure in the Napoleonic Era

What it felt like to be on the receiving end of such operations and the hammering which the landscape endured in those early years of the nineteenth century are painfully conveyed by another poet whose roots were in the East Midlands. By the Napoleonic era, when war with France intensified the need for food production, enclosure of common land by Act of Parliament, which in 1750 had generally begun to replace enclosure by agreement, was at its height. In 1809 an Act was passed for the enclosing of the parishes of Maxey and Helpston in Northamptonshire. One aspect of the landscape revolution which this entailed appears to have been major drainage works, which drastically modified the stream between the two villages, to create what is now known as the Maxey Cut. John Clare, in his poem 'Remembrances', describes the damage done to his parish by the axe of 'spoiler and self-interest':

O I never call to mind
Those pleasant names of places but I leave a sigh behind
While I see the little mouldiwarps hang sweeing to the wind
On the only aged willow that in all the field remains.

The 'mouldiwarps', or moles, are the bane of drainage men, since their tunnels play havoc with banks and channels; and even now, water authorities employ mole-catchers. Clare continues:

Inclosure like a Buonaparte let not a thing remain
It levelled every bush and tree and levelled every hill
And hung the moles for traitors—though the brook is running still
It runs a naked stream and chill.[42]

The solitary willow, the gutted brook: these were the things which Clare picked out as the climax of his catalogue of casualties in this, one of his finest poems. How absolutely it chimes with our modern experience of loss of a sense of place. In the mid-1980s, drainage contractors in the Midlands are still moving in on river valleys, starting with the stream itself and then clearing every adjacent hedge and copse as part of the same contract. The only differences which Clare would notice are that machines have replaced axes, and that the moles are now gibbeted on barbed-wire fencing.

Clare's contemporaries in the wetlands had reason to be concerned more for their own survival than that of moles and willow trees. In 1812 James Loch took over as Lord Stafford's agent at Trentham in Staffordshire. He was later to become notorious as the scourge of Sutherland for his role in the highland clearances. Rather nearer to his employer's family seat were the marshes of the Wealdmoors, just north of Telford and Wellington in Shropshire, which Loch set about draining with a will. Loch's reputation in Shropshire does not appear to have been so contentious as it was in Scotland, and the landscape he created in the Wealdmoors is now level ploughland of peat interspersed with rectangular plantations of poplar.

Moles, the bane of drainage men, gibbeted on a willow tree
as described in John Clare's poem 'Remembrances'.

The Somerset Levels. The men of the Levels successfully resisted the kind of large-scale drainage which transformed eastern England in the seventeenth century.

In the Somerset Levels, the surviving nucleus of wetland commons were tackled between 1770 and 1800. The period opened with the usual resistance from the inhabitants, who dug an open grave for William Fairchild, the surveyor of King's Sedgemoor, and announced 'a reward of a hogshead of cider . . . to anyone who could catch him'.[43] Nevertheless, by 1800 a commoner was speaking with regret of the times when the undrained wastes had given him pasture, where 'he could turn out his cow and pony, feed his flock of geese and keep his pig'.

## Otmoor

North-east of Oxford lies Otmoor, four square miles of damp land cradled in a basin of low hills and watered by the river Ray. It is ringed by villages known as the 'seven towns', which for centuries were sustained by the rights of common they enjoyed upon Otmoor's lush pastures. In 1815 an Act for the enclosure of Otmoor was passed, at the instigation of Lord Abingdon and George, duke

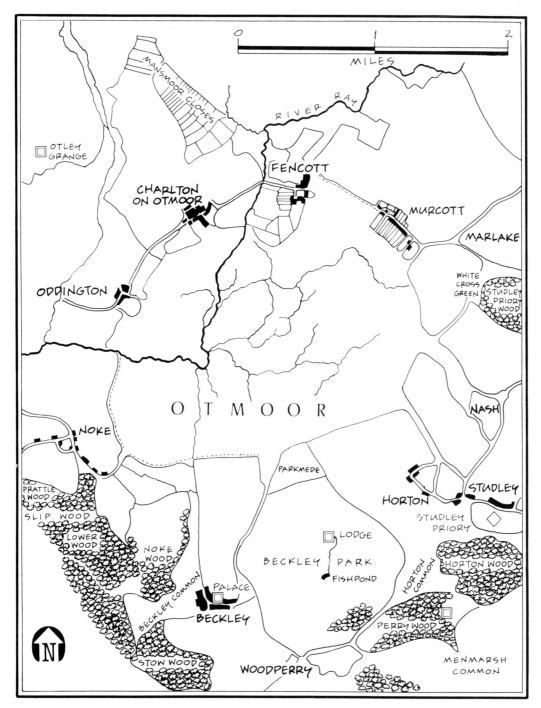

Otmoor and its seven towns before enclosure.

valleys, they sometimes disrupted the drainage, creating small marshes which survive to this day. A similar mixed result had attended the efforts of earlier canal engineers. Thus James Brindley set out to drain Longdon Marsh in Worcestershire, and the duke of Bridgewater organized drainage of the northern part of Chat Moss as part of his canal building. More often, however, navigation engineers were at odds with drainage men, especially when they were working on existing rivers. To make most rivers navigable, the water level must be raised by building weirs; there is little doubt that the navigation structures built on the

The great copper butterfly, now extinct through drainage, depended in its life cycle upon the water dock.

Leicestershire Soar in the 1770s worsened the local drainage, as did eighteenth- and nineteenth-century navigation works on the Thames and in Somerset.

In 1851, which, with symbolic appropriateness, was the year in which a hungry urban population exceeded for the first time the population of the countryside, drainage-minded landlords up from the shires were able to carry away a wealth of interesting ideas from the Great Exhibition. Pipe-tile-making machines were on show as was a new centrifugal pump which 'astonished the visitors'.[46] Within two years, Martin Mere, just south of the Fylde and Whittlesey Mere in the Fens, had been pumped dry.[47] By the late 1840s the English race of the great copper butterfly, which had retained its last stronghold around Whittlesey, had become extinct.[48] A symbol of the less controllable results of technology, the Holme Post, a cast-iron column believed by some to have come from the Crystal Palace, was sunk into the ground near Whittlesey, to measure peat shrinkage. Its replacement marker indicates a thirteen-foot drop between the present ground level and that of the mid-nineteenth century.

Technology was to transform not only the actual level of the land, but also the fine details of the wetland landscapes. Extracts from the account book of the Derwent Ings reveal the subtle alterations which took place in the space of two generations:

| 1867–8 | Oct. 17th | Leather for the Clow. |
| | Oct. 19th | Cheese and bread on Ings Breaking Day. |
| 1887–8 | | Paid J Binns for moles catching £1. 18s. 0d. |
| 1910–11 | | Taking one ton of cement to bridge. |
| 1923 | | Half cwt barbed wire fifteen shillings.[49] |

A 'clow' was a flap-valve to stop flood-water rising back up the ditches at high tide. Leather was no doubt obtained from cattle stocked on the Ings. Nowadays such things are made of steel and rubber.

William Morris. He challenged the Thames Conservancy over their felling of riverside willows. National Portrait Gallery.

The 1850s and 1860s, especially after the Land Drainage Act of 1861, saw a steady increase in mopping-up operations. In the 1860s Longdon Marsh in Worcestershire and the Baggy Moor in Shropshire were tackled, as were marshes in the Trent and Tame valleys in Staffordshire. In the 1850s the 'Bampton Polderland' on the Thames just south of Kelmscott was drained by an entrepreneurial farmer, William Wood. This still lovely riverine landscape must have been settling down from its major reshaping when William Morris bought Kelmscott Manor in 1871. Morris was no stranger to the environmental impact of river works, as is shown by a confrontation recorded between him and the Thames Conservancy Board concerning riverside willows which they had felled near Hammersmith. Morris resolved to complain to the board, which was composed largely of retired seamen. Feeling increasingly nervous as his appointment drew near, Morris was ushered into a vast boardroom, where the door was flung open by 'a giant of a man who looked as if he had been fed on rum all his life':

'What the hell do you bloody chaps want?' he roared.
'What is your bloody business?'
Morris' shyness disappeared in a flash. 'We've come to ask you savage bloody chaps why the hell you have cut down a pleasant grove of willows.'[50]

*The Twentieth Century*

Despite the decline in agriculture at the end of the century, the tree felling and river clearing continued apace. In the Great War, river clearance was seen as employment for prisoners, and in the 1920s as a way of relieving unemployment generally. In 1922 the chief drainage engineer for the Ministry of Agriculture reflected thus upon the constructive uses to which he had put the heroes home from the trenches:

> The real value of the results actually achieved has been a most gratifying surprise to everyone concerned, and it is easy, after the event, to reflect wisely upon the fact that a large percentage of the men must have served a long and painful apprenticeship, whilst on Military Service, to the art of transforming swamps into 'better 'oles'.[51]

To supplement the armies of the unemployed, new machines were becoming available. In 1920 the 'Buckeye' Ditcher, a bizarre-looking contraption of wheels and pulleys imported from America, arrived at the railway station at St Neots, and was escorted to Croxton Park, where a public demonstration of its capabilities was attended with considerable excitement.[52] Primitive drag-lines were soon adopted by the richer drainage boards.

None the less, the situation from a drainage point of view was far from satisfactory. Agriculture had been in steady decline since the 1870s, and after a brief post-war boom was to sink back into it again in the 1930s. Two centuries after the 'blowing' of the Denver sluice in 1713, the whole security of the south level of the Fens depended solely upon the security of the restored sluice. In 1919 there was little to distinguish the 'summer grounds' of large areas of the Somerset Levels from their saturated state a hundred years earlier. To compound the problem, administrative chaos reigned. There was everywhere 'a chaos of Authorities and an absence of authority'. In 1922 Sidney Webb wrote that the lords of the level of Romney Marsh, who had inspired the founding of the courts of sewers in 1258, still remained unreformed themselves, an ancient relic of pre-statutory local government.[53] The levying of 'scots' on Romney Marsh was discontinued only in 1932, after seven centuries of enforcement. Outworn institutions were to stagger on until, in some cases, the outworn machines and sluices which they maintained also collapsed. In 1940 the old Burwell drainage commission finally accepted voluntary dissolution when its ancient pump engine collapsed beyond repair. In 1929 the Somerset drainage commission went out of existence at the height of a great flood.

The Land Drainage Act of 1918 had made an attempt to simplify the administration of land drainage, but in the late 1920s it was clear that confusion still prevailed. In 1927 the Bledisloe Commission submitted a report which was to form the basis of the 1930 Land Drainage Act, the ground rules of which have dominated land drainage to the present day. Catchment boards were set up, which in 1948 became river boards, and in 1964 river authorities. In 1974 these were incorporated into the water authorities.

The Denver sluice. Washed out to sea in 1713 and successively rebuilt, it remains the key which controls the floods over the Great Level.

The still-vexed issue which the 1930 act attempted to resolve was who was to pay for drainage.[54] The courts of sewers, which were abolished in 1930, had always insisted on the principle: no benefit, no rates. The new proposals extended the rating for all works on what is now known as 'main river'—the larger watercourses—to the total catchment area. Internal drainage districts (which maintained smaller streams and ditches, especially in the lowlands) were to be funded from within their own areas. The lowlands thus required two sources of income for drainage: one to maintain the main river system, and the second for the back ditches. The householders who lived above the valley bottoms were not happy about paying two sets of rates, needless to say, especially when the towns on the tops of the hills found themselves subsidizing drainage benefits for lowland farmers. The argument put forward by the lowlanders, that they were being flooded out by water sent down to them by those living higher up the hill, was naturally resisted by the latter.

In 1933, following a typical dispute over this issue, Alban Dobson of the

the flood-banks in the Fens. Wentworth-Day, who lived through it all, describes the experience:

Throughout the black night came the dull thunder of the bursting banks, the village alarm of 'she've blowed'. The river is always feminine. In a thousand remote little farmhouses and cottages, islanded beneath wind-shriven willows or leaning poplars, the racing floods covered the black fields, overflowed the straight dykes . . . and leaping upon those lonely homes with all the relentless force of wind and gales, burst open the doors, shattered the ground floor windows . . . and rushed gurgling and swirling up the narrow staircase.[56]

In the freezing cold and pitch dark, families were driven to clinging to the roof. In the south level 37,000 acres went under water, and the chief engineer of the Great Ouse catchment board considered it the worst fen flood since the time of Vermuyden.

Disasters of this sort were confined to neither traditionally flooded land nor the winter months. In August 1952 the innocent-looking East and West Lyn rivers swept down into the Devon resort of Lynmouth, obliterating houses and removing all trace of the Beach Hotel, which was carried out to sea.[57] Thirty-three people were drowned, and the main street was transformed overnight into a dramatically boulder-strewn river bed. This flood, in a narrow valley, exacerbated, it was thought, by trees blocking the bridges upstream, was different in kind from the inundation of large areas of low-lying land in East Anglia. It helped set in motion renewed enthusiasm for tree clearance in upland catchment areas as yet another aspect of the land drainage solution.

The following February, floods again struck the east coast, from North-umberland to Kent. The casualty list was 307, and the devastation immense. On Canvey Island alone, 11,000 people were rendered homeless.[58] Coronation Day, in June 1953, was a memorably wet day. While the symbols of civilization processed with full panoply of State through the streets of London, out on the eastern marshes the North Sea nudged at the coastal defences which had been hastily shored up after the calamity just four months before. Our era was ushered in with a reminder that the flood remained not entirely tamed.

The twist in the story of river and wetland management in our own times was to be one in which drainage men were faced with a new army; an army which, unlike that of the commoners who had opposed their predecessors, was to grow stronger with every passing year. It was to come from an unexpected quarter: the environmentalists. Just as the drainers had previously arrived as outsiders on the wetland commons, so the conservationists were to wade into the wetland issue, often importing their perspectives from outside. All this was foreshadowed at the height of the war and in the heart of the Fens by the testimonies of two passionate and knowledgeable countrymen, one a bird-watcher, the other a farmer. Their conflicting attitudes to Adventurers' Fen, which the one painted and the other drained, are both completely understandable in human terms.

In 1942 Eric Ennion wrote *Adventurers Fen*, celebrating its birdlife and the

The power of a flood. A house collapses under the impact of water in the 1953 east-coast disaster. Photo: HMSO.

haunting beauty of its landscape. His celebration was also a requiem. He wrote:

It is more than a year since the red and white surveyors' poles glinted above the reeds, blazing a trail for the draglines that were soon to follow. They came, each with a gaunt arm cutting the gentle skyline, clanking and threatening, laying their tracks as they rumbled along. . . . In a few short weeks the scoops had torn a channel twenty feet wide from end to end, ripping the backbone out of Adventurers' Fen. . . . When all was dry, men set the fen on fire. Spurts of flame began to flicker here and there and presently leapt up to redden the fringes of the great smoke cloud which hung above them. . . . Reed beds, sedges and sallows vanished in a whirl of flying ashes amid the crackle and the roar.

I went down afterwards. There was a single gull wheeling over the black land and a wild duck trying to hide in two inches of water at the bottom of a drain. A couple of tractors stood waiting to begin.[59]

Alan Bloom, who had helped implement the scene described by Ennion, and was unusual in being both a practical man and a writer, published *The Farm in the Fen* in 1944. He conveys with a sure sense of atmosphere and detail the satisfaction of getting a job done, the feeling of being within a tradition of a long line of

Bog oak, the relic of wartime battles to drain Wicken Fen, remains stacked among the nettles in a corner of the fen.

settlers of the land, and the sheer hard work which that entails: from persuading recalcitrant committees to support his endeavours to the literally back-breaking labour of dragging obstinate lumps of bog oak out of the peat to facilitate ploughing. One morning, Bloom records, after 'a particularly bitter struggle with an oak', a Cambridge student came wheeling his bicycle up the drove-way. The young academic coolly eyed the embattled farmer and, commenting upon the destruction wrought upon the fen, looked forward to its return to wilderness after the war, so that it could act as a buffer for the nature reserve of Wicken Fen against the farmed land:

Against the farmed land, be damned, I thought, and let drive with all the most forceful arguments I could lay my tongue to. But I might have spared myself. When I had finished I could see that it had made not the slightest impression. We were in opposite camps, and he could no more appreciate my line of reasoning than I could see his point of view, and finally we parted.[60]

This conflict of values in which neither side has an absolute monopoly of truth has rumbled on, becoming louder and louder in the English countryside. In the latest episode, rivers and wetlands have often been the field of battle. The story of this conflict forms the subject-matter of the remaining chapters of this book.

'A wild duck trying to hide in two inches of water at the bottom of a drain', Eric Ennion, *Adventurers Fen*.

# 4

# The Wasting of the Waters
## The Real Cost of Orthodox River Management

IN 1977 the Ministry of Agriculture estimated that 6.4 million acres (2.6 million hectares) of agricultural land in England and Wales needed drainage.[1] This, approximately one-fifth of the nation's farmland, was expected to keep engineers in the newly formed water authorities fully occupied until well into the next century. In 1980–1, a fairly typical year, the grant from the Ministry of Agriculture for land-drainage schemes, including some urban work, amounted to about £30.5 million, of which £23.7 million was allocated to the water authorities, while the rest went to internal drainage boards and local authorities.[2] In that year it is estimated that the water authorities spent around £40 million on capital drainage schemes for agriculture. In addition, all water authorities spent large sums on routine river maintenance on farmland; and still more money was spent by the Ministry of Agriculture in contributing between 30 and 70 per cent of the cost of farmers' field drainage, designed to pick up the benefits of lowering the water level, a result of river engineering schemes.

In the winter of 1984–5 the Government announced a cut in the annual grant for land drainage from £60 million to a projected average of £30 million, and stipulated that the emphasis should now be on urban flood protection, rather than improvement for agriculture. None the less, large sums of public money are still being spent on drainage, much of it to improve farmland. In the financial year 1986–7, the Severn–Trent Water Authority spent £14.2 million on land drainage, of which £7.6 million consisted of river maintenance work for the benefit of agriculture. (Even so, it was still carrying out less than a fifth as many maintenance dredging programmes as the North-West and Thames Water Authorities.) Of Severn–Trent's three large capital schemes in 1986–7, two were almost entirely agricultural. In the same year, most of the county councils within the Severn–Trent region were carrying out some agricultural land-drainage work, and private farmers, even without drainage grants, were bringing in dredging machinery on rivers, to get water off their land.

## Agricultural Reasons for Drainage

Much of this expenditure, which has taken such a toll, on both the exchequer and the environment, has been based on a simple assumption which is deeply embedded in agricultural thinking: that excess water is the enemy of good husbandry and must be driven off the land. In those places where stock is kept out in winter, undrained grassland is unusable during floods, and in summer is susceptible to trampling by animals, and so supports less stock than drained pasture. In addition, damp conditions encourage husk in cattle and fluke in sheep. For the cereal farmer, wet land is late land, since waterlogging keeps the ground cold and holds back germination. Saturated soil lacks air, without which plant roots die, helpful micro-organisms are checked, and the soil loses its structure. Roots of wheat which have been stunted by excessive water in spring will be less able to withstand a summer drought. Actual flooding can be disastrous. Depending on the time of a flood, root crops can be completely destroyed if submerged for more than twenty-four hours.

## Agricultural Over-Production

The real problems which undrained land presents to farmers have justified drainage during periods when we have been short of food. What is now questionable is the national need to produce a maximum, perfect crop on every piece of English farmland. The process whereby rivers have been straightened and lowered to allow all riverside land to be more intensively cropped for grass and grain has often been expensive in terms of both wasted investment and loss of landscape. The environmental cost has been the destruction of wetlands and the general erosion of the rural landscape, whereby hedges, woods, and ponds have been removed. The habitats which have been hardest hit by the post-war agricultural revolution are on land which was always regarded as the most marginal: the barest hilltops, the steepest hillsides, and the wettest valley bottoms. The reason why these had survived so long was because they required the most money spent on them—in the case of valley bottoms, to pay for initial drainage and then to maintain it—to make them yield their full potential of arable and pasture. Now, with the massive harvests of the 1980s, when the superabundant corn bows its head along the banks of our chastened, canalized rivers, many of those who set out to tame the flood have succeeded beyond their wildest dreams. All over Europe there lingers the sweet smell of excess. In the south of the Continent, piles of surplus peaches and tomatoes are bulldozed, and cauliflowers are mixed with cod-liver oil before being buried. In the north, agricultural officials no longer discuss whether the harvest was adequate, only how the latest addition to the grain mountain can be stored. In 1985, as reported in *The Times,* the cost of simply storing the United Kingdom's cereal surplus amounted to around £111 million. That same year, the agriculture commissioner for the

With the massive harvests of the 1980s, cereal crops bow their heads over our chastened and canalized rivers.

Common Market revealed a new solution for reducing the butter mountain: feeding it back to the cows.[3]

Such is the latest outcome of forty years of enthusiastic and single-minded agriculture, of which an equally enthusiastic drainage policy has been an essential part, that conservationists are now questioning the aims of intensive river management on its own terms of hard-headed economics and efficiency. The wetlands, long regarded as wastes by generations of farmers, have been replaced by a harvest which fits the dictionary definition of 'waste' in every sense. Our annual surplus of grain is roughly equivalent to the annual yield of the Cambridgeshire fens, won from the flood by Cornelius Vermuyden 300 years ago. Intensive cultivation and continued drainage of the Fens further accelerates the degradation of the land, which is increasingly subject to peat wastage through oxidation and windblow.

Recently there have been determined efforts to reduce our food surpluses, but we are only starting to learn that river and land management require careful thought before instant expenditure of money. The drain-all, strip-out approach to land has been adopted by insurance firms investing in agriculture precisely because it requires no more thought than it takes to fill in a form for the subsidy. The practical reason why it is wise in the long term to give much more careful thought to river and wetland management is that drainage can contain, profoundly, the seeds of its own destruction.

## Physical Destruction from River Engineering

The real cost of river mismanagement can begin with engineering works on the watercourse itself. As many of the old river managers learned from first-hand experience of tinkering with their rivers, it is ultimately far more productive to work *with* a river than *against* it. It is the nature of rivers that they refuse to stay straight. 'It's always twisted,' said a seasoned riverman of the Shropshire Rea, 'and it always will.' Cage a river in cement and iron, and it will struggle to break out like a wild beast. Major straightening of the Mississippi in the 1930s, largely for navigation purposes, is still creating problems for its present-day managers over hundreds of miles. Attempts to straighten out the Lang Lang river in Australia between 1920 and 1923 caused a series of cuts into the bank, which progressed rapidly upstream and destroyed seven bridges.[4]

But it is not necessary to look so far afield or so long ago for examples of rivers which have refused to obey the dictates of engineers. In 1864 the river Ystwyth in Wales was straightened to run parallel to a railway track. In 1969 it was back again on its wandering course, and was engineered cheaply back into line by dredging out the gravel shoals which were causing it to wander. However, this did not turn out to be such a cheap solution in the long run, since it set up conditions of even greater instability, necessitating repeated operations every two years.[5] Major work on the river Taff and the river Usk in South Wales, carried out in the early 1980s, has precipitated extensive and unforeseen repair bills. Those who win money on the horses at Kirkby Lonsdale races in Cumbria can quickly sober up if they wander down from the racecourse to see how their rates have been spent on the adjacent reaches of the river Lune, where the recently cemented banks are dramatically caving in, and attempts at bank protection appear to have made matters worse.

The river Trannon is a fast Welsh mountain stream, which flows down towards the Severn in Powys. In 1978–9 a short reach near the village of Trefeglwys was given a thorough canalizing treatment. Trees were cleared from the banks, and raised flood-banks were built out of the dredged material alongside the stream course. Almost immediately it became clear that the river was kicking back at the abuse it was receiving. In the winter of 1979–80 the Trannon careered off on a course of its own. Fencing, which had been set well back from the bank, now dangled over the gulf created by the all-swallowing river. As more and more stone-filled gabions were built in to reduce the erosion which the scheme had set in train, it became apparent that Pandora's box had been opened, and that what might have worked as a piece of traditional river canalization in the cohesive sediments found downstream had proved a recipe for disaster when applied in the unstable gravels of this upland brook. At one point, the original river bank, shored up by wire and stone, remained the only still point, actually down the *centre* of the fast-moving Trannon, so that it was a job to guess whether the buckling gabions had belonged originally to the left or to the right. Seven years later, remedial works were still being carried out; the raised flood-banks set far too close to the watercourse were being eroded away; the stability of the downstream bridge was threatened; and the real cost to the public purse of what was a relatively small scheme originally has yet to be clearly counted. In 1982 the Institute of Hydrology carried out trials on the Trannon, and in 1986 was able to come up with a number of constructive lessons to be learned from this sorry story. Relatively cheap methods of testing local ground conditions which have been researched by geomorphologists can act as useful warnings to engineers as to whether they are risking the kind of problems which now make the Trannon scheme, with hindsight, a questionable one to have undertaken in the first place.[6]

Working with nature is clearly practical, as well as ecologically sound. In 1985 dredgers were busily raking up gravel from the shoals in the river to put into more gabions for bank reinforcement beside the Trannon. As part of the normal processes of a river, the stones in these shoals are neatly sorted and graded by the flowing water into an overlapping fish-scale pattern known as 'armouring', which makes them relatively stable. By dredging up gravel, therefore, river managers were actually de-stabilizing the river bed, thereby contributing to the erosion which they were supposedly trying to prevent. Riverside trees, whose roots increase the tensional strength of the bank material, are the best protection against erosion, especially if they are well established and are properly maintained. They also reduce land loss, since it is estimated that channels with

An engineered river moves out of control. After the trees had been removed the stone and wire were set to reinforce the bank. The stream then split, turning the old bank into an island. River Trannon, Powys. Photo: Malcolm Newson.

50 per cent tree and shrub cover on both banks require only approximately half the width for a given volume of bankfull flood-water speeding through the channel, compared to treeless brooks which erode out into the adjacent fields.[7] If trees are felled in the hope of gaining extra land, a river is likely to move out to take that land, and on certain types of river a great deal more besides. One has only to stand on the bridge over the Trannon at Trefeglwys and look upstream to see the stable narrow river coursing elegantly between its magnificent borders of ash and sycamore, and compare this with the immediate downstream reach, which wanders amidst a waste of gravel.

*Nitrates*

Additional benefits of tree-lined streams, in reducing the effects of nitrates and phosphates in the water and in shading out the choking growth of summer weed in a river bed, which may otherwise necessitate further expensive dredging, have been recognized by Dutch and German scientists for over a decade.[8] The only green thing about many of our English rivers nowadays is profuse algal growth in the water, which results from an excess of fertilizer leaching off arable fields straight into the stream. Traditional drainage has accelerated this problem in two ways: by promoting farming methods which require high levels of fertilizer, and by stripping out the buffer of vegetation between the riverside and the fields. It is not uncommon for tractors to slide into a stream as they attempt to cultivate every last scrap of land beside a watercourse. This approach has led to over-exploitation of stream systems. Until 1986, the steeply graded banks of one river in arable country were regularly sprayed with the approved chemical 2-4-D amine, in order to reduce the nettles, which were less good at holding the banks than slow-growing grasses, but for which perfect conditions had been established by abundant nitrogen leaching off the fields. As is so often the case in operations which go wrong, over-heavy management went hand in hand with a haphazard lack of supervision. The spraying boat, affectionately known as the 'Black Pig', would set off on its regular journey up the river without any adjustment on the spraying nozzle. Consequently, once the motor was started, the spray was meted out indiscriminately to riverside walls, fishermen, lovers on the bank, and nettles alike.

It is estimated that for every pound a farmer spends on fertilizer on his fields, he can expect the rain to wash away a good fifty pence-worth. But the real cost of such waste is not only monetary. Water is extracted for drinking purposes from many of our lowland rivers, and high nitrate levels in drinking-water can lead to illness in bottle-fed babies. Thus, while land-drainage departments promote agricultural intensification, other departments in the same water authority have to spend more public money in expensive plant to treat the water, in order to comply with EEC regulations on nitrate levels. Further expense to the water authority in cleaning up river pollution caused by silage effluent is yet

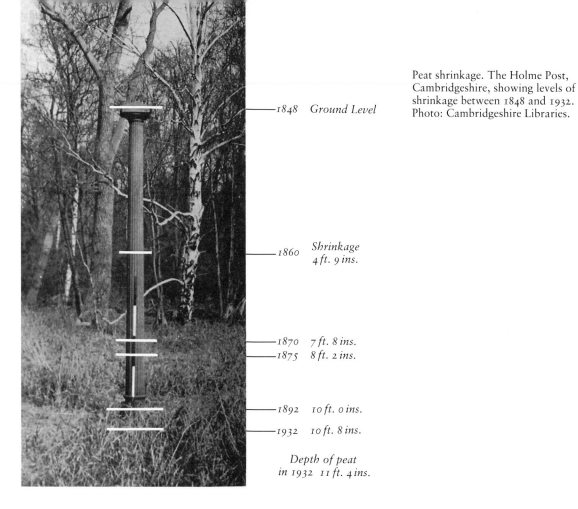

| 1848 | Ground Level |
| 1860 | Shrinkage 4 ft. 9 ins. |
| 1870 | 7 ft. 8 ins. |
| 1875 | 8 ft. 2 ins. |
| 1892 | 10 ft. 0 ins. |
| 1932 | 10 ft. 8 ins. |

Depth of peat
in 1932  11 ft. 4 ins.

Peat shrinkage. The Holme Post, Cambridgeshire, showing levels of shrinkage between 1848 and 1932. Photo: Cambridgeshire Libraries.

another consequence of high-gear monocultural farming, encouraged by land drainage.[9]

## Peat Shrinkage from Drainage

So much for the rivers themselves. Enthusiastic attempts to drain wetlands throughout the 1970s and early 1980s have in many cases failed to produce the high-quality farmland which was the object of these expensive exercises. It has now become increasingly clear that one of the real harvests of wetland drainage can be the physical ruin of the land. The most familiar example of such degradation is peat wastage. Peat wastes as a result of shrinkage, oxidation, and bacterial action, all triggered by the drying effect of drainage upon peat. A ten-year trial by the Ministry of Agriculture, started in 1980 on the Norfolk peat marshes near Acle, is indicating a fall in ground level of an inch each year. The level of the peat fenland in Cambridgeshire has fallen by up to 4.6 metres in many places since Vermuyden's initial scheme was completed in the 1650s. The Holme Post, recording a drop of 3.9 metres since 1850, presents an unrealistically optimistic picture of the problem, since it now stands in a damp nature reserve.

More convincing estimates of wastage can be made by assessing the height at which roads and even some wartime pillboxes stand proudly above the adjacent level of the fens. As the peat wastes, drainage of the land deteriorates, and so the drains are deepened, thus triggering further wastage. The lowest deposits of peat are also the most acid and so least valuable to the farmer. The Fens now resemble a gigantic and very profitable grow-bag. Like a grow-bag, however, they cannot go on producing their rich harvests of vegetable crops such as celery and carrots for ever, since once the peat has all wasted away, poor acid subsoils, especially clays, are often all that remain beneath. It is estimated that by the first decade of next century, only 20 per cent of the peat soils now present on the 561 square kilometres marked by the soil map of the Ely district will remain. Once the peat has gone in such places, some land may be insufficiently fertile even for growing potatoes, and the very best land in the country will have been reduced to mineral soils of only average quality. Such downgraded land will still have to be pumped and embanked, of course, since it will have dropped even further below sea-level,[10] a situation worsened by the fact that, due to the tilt of the land, eastern England is slowly but steadily falling in relation to the level of the North Sea.

## Erosion

Land drainage has also accelerated wind erosion, not only in the East Anglian fens, but also on the fine sands and silts of Lancashire, the Dee estuary, the Nottinghamshire carrs, and the Vale of York. In such places, ditches, which have a critical job to do in carrying away flood-water, are often filled up to the top with loose soil, following a 'blow'. Water erosion, whereby rain washes the topsoil off hillsides, is becoming an increasing problem in the United Kingdom, worsened by high-gear agriculture and drainage. It is especially serious on land bearing winter cereals, since the resulting ploughland can be exposed to the elements from October to April. Hedges and ditches, especially those separating a ridge crest from a valley side or channelling water safely along a valley floor, are often in critical positions to stop runoff. Once they are removed, all that soil, like the fertilizer, has to go somewhere, and, often as not, it ends up in the watercourse, silting up the stream bed, eroding banks, and worsening flooding. Thus drainage schemes, carried out for the express purpose of intensifying agriculture, converting land from grass to cereals, and 'rationalizing' hedge and ditch systems, may create further drainage problems in their turn.[11] When the machines move in a second time to desilt the river bed, they will of course send the silt further down the river for yet more diggers to tackle downstream. Flooding, affecting whole villages, as a result of soil runoff has taken place in Shepton Beauchamp in Somerset and Ashford-in-the-Water in Derbyshire. In 1983 Lewes District Council spent £12,000 to alleviate the effects of soil erosion on a housing estate.

Attenborough withy beds. *Top.* In March the different withies intensify in colour. *Bottom.* Willow used as an instant fencing material.

'The Air nebulous ... the water putred and muddy, yea full of loathsome vermine; the earth spuing, unfast and boggie.' Our ancestors' reaction to the wetlands was one of horror.

## Acid Sulphate Problems from Drainage

On low-lying land, drainage can create another major problem, one associated with acid sulphate soils.[12] In ditches and damp springs, you may sometimes notice that the water is discoloured to a strong shade of orange. This is caused by 'iron ochre', a generic term used to describe the iron-derived material which blocks drains. Iron sulphides occur naturally in the form of mineral iron pyrites, derived from rock slate or else found in peat and marine deposits. As drainage exposes these sulphides to oxygen, they oxidize, producing sulphuric acid and ferrous sulphate, which flows towards the drains. As a result of the increasing acidity, certain acidophilic soil bacteria are then released, which promote further acidification. Two problems arise for the farmer: first, ferrous iron creates an iron pan, blocking soil fissures and root channels, and, worst of all, sealing pipes and drains; second, the acidity of the soil reduces crop yields, especially of cereals. The treatment for both problems is expensive, and has to be repeated constantly, thus rendering the initial investment in drainage quite worthless in some cases, and at other times leading to a downgrading of the agricultural classification of the land. The cost of liming acid soils is phenomenal, as is the continual jetting and rodding of clogged pipes. Some land at the south-west edge of the Fens has recently been downgraded from grade 1 to grade 3 because of its acidity and the presence of bog oaks which have been exposed by wastage.[13] Acidity and ochre formation, triggered by the lowering of the water-table, are not confined to traditional wetlands. They are well known in the Weald of Kent and Sussex and around Poole harbour; and one land-drainage contractor was of the opinion in 1986 that in his own county, Northamptonshire, there was scarcely a parish without areas which were effectively undrainable due to ochre.

## Flocculation Problems from Drainage

Coastal marshes, the controversial targets of environmentally destructive drainage proposals throughout the 1970s and 1980s, are also proving in many places to be simply undrainable, since, after five to ten years, soil can be subject to a process whereby the chemical bonding in the clay structure breaks down. This is known as 'flocculation', and is especially problematic on drained estuarine soils with high salinity.[14] With increased drainage and ploughing, drain-pipes become clogged with clay and silt, and surface ponding creates ever more difficult conditions for farming, and even the death of cereal crops. This has been a major problem in Kent, Essex, Suffolk, and Norfolk, and most notoriously at Halvergate, where there have also been acidity problems, causing some farmers to have to jet their drains every year. In 1986 *Farmers' Weekly* described the difficulties now overtaking land on the Isle of Grain in north Kent, where permanent pasture was drained and ploughed in the 1960s and early 1970s. For a while, profits on this grade 1 land came rolling in, but ten years after the initial drainage, the worst areas required re-draining, and even high-pressure jetting of clogged drains was futile against the deflocculated clay slurry. The soil surface resembled 'thick Windsor soup'.[15] One solution increasingly adopted by farmers has been to return the land to grass again. But the new grassland is less environmentally interesting than the previous permanent pasture, and the money wasted on the whole ten-year cycle of grass-grain-grass has been very considerable.

### Professional Fees

The way in which professions operate within the construction industry also tends to favour expensive engineering schemes on rivers, rather than small-scale operations. In the public sector, it is inevitable that certain engineers will promote as much work as possible, in order to keep their empires afloat. An expensive all-concrete, all-dredge, work-from-both-sides approach may in certain cases have much to recommend it to a few of the more old-fashioned river managers, especially since the cutting of straight ditches involves minimum initial design and subsequent supervision, combined with maximum actual earth-moving and construction works on the ground. The salaries and career prospects of in-house engineers are linked to performance aims, which can encourage over-design of schemes. The fees of consulting engineers from outside, not to mention consulting environmentalists such as landscape architects, are generally based on a percentage of the total construction contract. The larger the job on the ground, the greater the professional fee to be made out of it. In fairness, it must be said that the professions have worked out a number of checks and balances to this system, and a common alternative to a percentage fee, hourly rates, can work out as an even more expensive open cheque drawn on the public purse. In addition, one of the duties of a water authority engineer controlling an outside contract is to tie consultants to a tight brief, which specifies the minimum work required to achieve the necessary flood reduction, while at the same time protecting the environment. None the less, the subtly corruptive potential of a percentage fee system, especially in our enterprise culture, is not hard to imagine.

Perhaps the most startling case of a professional fee system which taxes both the public purse and the environment is that by which land-agents charge for their services on land-drainage schemes. Land-agents are employed to act for the landowner in negotiating compensation for disturbance created by the working space needed for engineering operations, which may sterilize crucial areas of corn or pasture for a season or more. The landowner's agent may also act for him in negotiations with conservationists and engineers to secure the parcels of riverside land required for habitat protection and enhancement. Although some remarkably conservation-conscious agents exist, it is generally true to say that land-agents have played, and continue to play, a significant role in hastening the destruction of the English landscape by intensive agriculture. Farmers, while including all gradations of responsiveness within their ranks, do in many cases have a deep feeling for the land which they see every day, and which in many cases was shaped by their fathers and grandfathers, and which they would like to pass on to their children. Land-agents and estate agents passing through at the brief moment of negotiation are likely to be more exclusively dominated by the money motive. Most people who negotiate for habitat in the countryside will bear witness that their hearts sink when an agent is involved. Time and again a landowner, the man who actually has the most to lose since it is his land, is willing to donate corners of his fields for ponds or tree planting, as his contribution to the environmental compromise of a land-drainage scheme; but then a letter from his agent arrives, demanding that these corners be heavily compensated or even bought by the water authority, as payment for the concessions made. A typical compensation sum demanded by an agent for a half-acre corner of land given for a pond or tree planting might be £750, taking the market value of grade 2 land in the Midlands in 1986. Considerable hidden costs in time taken by authority estates and environmental staff over negotiation could be added to this bill. Riverside lands on which attempts at habitat creation by engineers and environmentalists have been abandoned are more often than not those where the demands of agents have thwarted such proposals.

One look at the fee scale by which agents are paid quickly explains why this situation exists. The scale set by the Inland Revenue and the law for agents dealing with public authorities is known as Ryde's Scale, and was amended in 1984 by the 'Revised Annex to Ryde's Scale'.[16] The fee charged by the agent is, naturally enough, directly proportional to the compensation he gleans from the public authority on behalf of his client. Thus, at figures current in 1986, if the cheque received by the landowner for compensation is up to £100, the cheque sent to his agent will be £85. If the landowner's compensation amounts to £200, or even just tips over to £101, then his agent will notch up to £170. Consequently, at the bottom of the scale, the agent may collect more money than his client in compensation. The public, as ratepayers, should realize that not only is all compensation paid out of the land-drainage budget, but that the fees of the landowner's agent are also financed entirely out of the public purse of the water authority or other public body carrying out the scheme. Therefore, in those cases where land-agents levy heavy penalties for such things as tree planting (which no engineer or environmentalist would attempt to force upon farmers in places where they would interfere with necessary practical farming), the real cost to the public, which is, after all, financing land-drainage schemes in order to improve the productivity of landowners' fields, is in both money and loss of landscape quality.

## Unsupervised River Maintenance

The possession of large machines by those responsible for managing rivers can sometimes lead to a situation where the tail wags the dog, and the results can be damaging to both the environment and the economy. In real life, a degree of flexibility is inevitable for the river managers, despite the essential submission of their work programmes to environmentalists well before the diggers arrive on the river bank. Farmers change their minds about access to a river at the last minute, according to the vagaries of the harvest. Altering weather conditions can accelerate or slow down dredging programmes, and the engineer for a direct works section must also keep his construction plant and men employed in order to make them viable. When something goes wrong, the slogan 'Have machine, will travel' has a depressing ring of truth, especially disastrous in the winter season, since it is no good excavating a channel in midwinter if the ecologists working with the engineers on a scheme have not had the opportunity during the previous summer growing season to identify the existing aquatic habitat affected. Such problems of inadequate pre-planning arise most commonly in river maintenance work, which is indeed the Cinderella of land drainage.

All engineering on rivers in England and Wales falls into one or other of two categories: capital schemes and maintenance schemes. At first glance the distinction seems clear. Capital schemes, which are partially funded by grants from the Ministry of Agriculture, make substantial drainage improvements; whereas maintenance schemes, paid for by rating precepts, aim at stopping floods from getting worse, by removing blockages such as fallen trees and shoals. However, since maintenance schemes often involve dredging and considerable tree removal, their impact on river habitat makes nonsense of the long-cherished claim by some river engineers that river maintenance is too minor an affair to concern environmentalists. The distinction between capital and maintenance schemes can sometimes amount to little more than an administrative technicality, and as grants for capital schemes are reduced, there is a possibility that old-style capital works will be slipped by under the banner of maintenance. In the financial year 1984–5 the North-West Water Authority carried out capital-works dredging on 21 miles of river in its care. In the same year, under a maintenance budget, it carried out full-scale dredging on 279 miles of river. That same year, the total length of river dredged under capital schemes by water authorities in England and Wales amounted to an estimated 223 miles, as opposed to 1,165 miles of maintenance dredging.[17]

This trend presents anxieties to environmentalists since maintenance schemes have traditionally escaped not only all cost–benefit assessment, but also, in many cases, the careful integration of conservation principles. In the remoter corners of water authority empires, which have had little more than a decade in which to professionalize themselves since their formation in 1974, gangs of river maintenance staff inherited from the far less environmentally accountable river boards have guarded their independence from interference by senior central management within their own organizations.

In 1954 the Ministry of Agriculture's chief drainage engineer gave a paper on land drainage to the Institute of Civil Engineers. As reported in the proceedings of the institute, this was followed by some comments on river maintenance which might have been made only yesterday:

Engineers naturally preferred spectacular schemes so, being human, they tended to concentrate on new and important works, leaving maintenance to the foreman. . . .

If the foreman saw a tree in a channel, or a shoal forming, he did not worry whether it was doing any harm or not; he played safe and had it out. Consequently, unless carried out carefully and under constant supervision, maintenance tended to cost more than it should.[18]

Not only have such unsupervised activities been questionable in terms of economics, they have also had a profound effect on the river environment. Pressure from conservationists, so often accused of adding a burden of unrealistic expense and inefficiency to practical operations in the countryside, has frequently provided the impetus for more efficient supervision of these roaming river-maintenance gangs. In the past, and sometimes even in the present, bad environmental management of rivers has been only a part and a symptom of bad man-management. If you cast a shrewd eye around the countryside in March, you cannot fail to notice that spring is heralded not only by splashes of yellow celandines and pale primroses in the hedge-banks, but also by gleaming yellow diggers, hastily working their way up the watercourses to use up their allocated budget before the financial year ends on 1 April.

## Cost Benefits

Such are some of the real costs of taming the flood in our times: damage to the environment, over-production of food, and a decline in some places of the agricultural quality of the land. It would be desirable under such circumstances if the Ministry of Agriculture could be persuaded to produce a proper definition of farming practice which takes into account the interests of good, long-term husbandry. Throughout the 1970s and 1980s, controversy has raged over cost–benefit studies carried out by water authorities to justify large land-drainage schemes. Conservationists have levelled a series of criticisms at the basic ground rules of these cost–benefit studies.[19]

These criticisms have often been the crux of the public case against land drainage. Environmentalists have complained that in contrast to road-building programmes, for example, no attempt has been made to assess the cost of intangible environmental losses to the community, as against financial benefits to the farmer; that calculations of benefit have assumed unrealistic yields and excessively speedy rates of take-up by farmers; that there is a reluctance to design low-level flood protection, even when farmers are getting by with an arable crop in most years; that the inevitable patching of eroding banks as a river reacts to the engineering constraints put upon it is never allowed for in the costs; and that the benefits anticipated from drainage schemes are based on what are known as

A combine-harvester gathers in its crop. In 1985 the cost of simply storing our cereal surplus was around £111 million.

'farm-gate prices' received by farmers for their crops. These prices are un-realistically high because of both the protection which agriculture enjoys and further hidden subsidies. Thus the lack of balance which strikes anyone looking at a thoroughly drained, treeless landscape is reflected in a lack of honest balancing in the books and ledgers which lie behind that landscape. Lastly, the opponents of drainage schemes have long complained about the secrecy with which the Ministry of Agriculture and the water authorities have shrouded their calculations, together with their almost universal resistance to public inquiries, which are now a normal part of the process of consultation preceding road and reservoir schemes.

The argument for secrecy has rested on claims that it is unreasonable to reveal landowners' financial affairs, but such discretion seems unjust when large amounts of public money are being handed out for the benefit of these landowners, and when the financial data central to the justification of schemes need not involve revelation of every last penny of landowners' private accounts. For a long time, the secrecy maintained by protagonists of drainage has given the impression that they have something to hide. In 1985, however, the Ministry of Agriculture took a first step towards meeting environmentalists' objections, with a new 'Guidance for Drainage Authorities' on investment appraisal.[20] Environmental intangibles have been built into the cost–benefit analysis in the same way as they are for road schemes. The ministry accepted that farm-gate prices were a misleading basis for calculating cost benefits, but the new methods they came up with did not fully adjust the calculations accordingly, especially in the case of cereals. On the point of secrecy, some information was at last released, but with insufficient hard financial data for opponents of schemes to interpret it clearly.

# 5

# Riverside Riches
## The Need for Management on Wet Land and some Alternative Economic Uses

IN the reasonable belief that the production of surplus food cannot be allowed to continue for ever, farmers are starting to look for alternatives to high-input agriculture as sources of future income from their land. Diversity has become a watchword throughout the countryside, not least on land which was previously considered a prime candidate for drainage. Those concerned with the quality of the landscape are also beginning to look in the same direction as those who must make their living from it. The beauty of the traditional English countryside has been that it was a working landscape. Hedges, trees, and damp pasture did not impede practical farming, and often contributed directly towards it. Many valleys in which there was room for willows, flowery pastures, and river margins of rush and reed alongside decent fields of corn have now been converted into monocultures of grain or rye-grass leys, which must be regularly ploughed and resown to sustain productivity. To maintain this high level of production, for which a satisfactory market is no longer assured, a high input of effort and expense will have to be kept up. Even in the summer of 1986, when there was no excessive early rain to spoil a perfect harvest, flash flooding damaged the corn which some farmers had sown right down to the river banks. The flood is never permanently tamed.

Academic economists, assessing the cost benefit of drainage schemes, have long been teased by the problem of putting a monetary value on the spiritual and aesthetic qualities of landscapes affected by such operations. It would be absurd to insist that flowers and birds should be permitted to exist only if they pay their way; but it is now possible to look closely at our riverside landscapes and ask which is the harvest and which the waste, which is the weed and which the crop. In this way we can perhaps move towards formulating an economy for the ecology, and a set of values on which to base wise management of our rivers and wetlands.

*Wetland Management: Wicken Fen*

Management, rather than total neglect, is very much what conservationists want for the lowland landscape. To understand the extent to which ecologists manipulate the landscapes in their care in order to achieve diversity of wildlife, a good place to start is Wicken Fen, the oldest, and perhaps the most famous, wetland nature reserve in England. Its 605 acres (245 hectares) have probably been more studied, more written about, and more experimented with than any other parcel of land in the country. To speak of Wicken is to speak of the origins of ecology as an academic discipline, for it was to Wicken Fen that Cambridge ecologists such as Tansley and Godwin went to study relationships among different plant communities. The earliest plot to be bought as a nature reserve was acquired by the National Trust in 1899. Wicken is a remnant of the natural undrained Fens, surrounded by intensive Cambridgeshire farmland. But it is scarcely a wilderness, for all its wild appearance. Adventurers' Fen, which now forms part of the reserve, has been cultivated intermittently since the Merchant Adventurers drained it in 1656. The rest of the site has a long pedigree of practical management. Peat has been dug there at least since the seventeenth century, while purple moor grass was mown for cattle bedding, and saw sedge was harvested, as it still is, for the local thatching trade. Houses in the village of Wicken are roofed with reed and sedge from the fen, and their walls are built of brown bricks like those which must have come out of the fen brick pits operating during the last thirty years of the nineteenth century. The natural environment of the reserve has benefited from its past contribution to the local architecture. Sedge cutting is essential to maintain the botanical diversity of the site. The brick pits, now as old as the century, have been colonized along their margins by both species of reed-mace, and their cool waters reflect the perfect flowers of the rare white water-lily.

A visitor's first impression of Wicken Fen is of a wet wood. Woodland rides, bordered by yellow iris and the blue-green spears of reed, veer off into the dim green heart of the thicket. The tawny sedge-beds, closely bounded by a dense scrub of ragged birch, have a remote, almost African feel about them. They suggest a clearing in the ancient wildwood, through which one can imagine Plantagenet kings riding in pursuit of wild boar. Such forest covered all of lowland England before man's arrival, and without us it would certainly return. Ecologists manage Wicken so as to deflect and delay the advance of this wood. On a series of plots spread over the reserve, every stage in the succession between total clearance and maximum tree cover is held, as if frozen in time, through a strenuous and sophisticated programme of mowing and cutting.

Look more closely at the fen, and you will see that it resembles a green chequer-board, dead level and gridded by intersecting dykes and droves. This is a characteristic of many wetlands, as Lewis Carroll appreciated, after visiting Otmoor, when he invented the chess-board landscape in *Alice Through the Looking Glass*. Wicken is a chequer of many shades of green: the tenderest

The cool waters of a brickpit in Wicken Fen are now colonized with white waterlilies.

emerald reed, gold-green sedge, and the different shades of willow and water. It is also a chequer-board in three dimensions. The vegetation on the various plots, some of them relating to the strips allotted to the commoners of Wicken in 1663, ranges in height from carpets of delicate moss, through grass meadows, waist-high sedge, tall reed-beds, and scrub thickets known as 'carrs', to the full height of mature woodland, towered along some drovesides by the graceful birch. Each of these stages of fen vegetation supports a different plant community. The grass fields are flowery with comfrey and hemp agrimony. The sedge is mixed with milk-parsley. The reed has a specialized bird-life of harriers and bearded tit. Alder buckthorn in the scrub ensures the survival of brimstone butterflies, and the woodland provides a home for the sparrow-hawk, woodcock, and long-eared owl. Even the droves, or pathways, vary from close-mown to foot-high swards, pink with orchids and ragged robin.

Sparrow-hawk.

Thus, every permutation of the natural fen vegetation is created by a far from natural process of management which has its origins in the old economic harvests of the fen. The result is a landscape which is unified and yet infinitely varied: a chequer-board garden, dappled by the cloud pattern of the Cambridge sky, fragrant with mint and the rotten-sweet river smell, and loud with the scraping chatter of sedge-warblers and the insistent cuckoo. Yet it has taken almost a century of refining Wicken's management to formulate the best regime for mowing the sedge. This was previously cut until as late as October, thereby exposing up to 50 per cent of it to be killed by frost. Now, by completing mowing by the end of July, it is hoped that the flora of the sedge-beds, including milk-parsley, which feeds the swallow-tail butterfly, will recover its full vigour. A detailed study of Wicken's history completed in 1983 showed that an early summer cut was traditionally adopted by the commoners.[1] Thus, scientists have finally come to the same conclusion reached by those for whom the sedge had provided an essential livelihood. It is hoped that this early cut, combined with increased saturation of the fen as a result of bank repair by the water authority, may eventually create favourable conditions for the reintroduction of the swallow-tails, absent from Wicken since 1951. Only by such careful management will the black and yellow butterflies for which the fen is famous return to their ancient haunts.

## Management of Ditches for Wildlife

The dykes at Wicken, which currently support seventeen species of dragonfly as well as water violet and yellow water-lily, depend for their natural interest on a rotational winter dredge to prevent their choking up. This is true of all ditches in the wetlands, which, after all, were dug in the first place for the practical

Woodland rides, bordered with the blue-green spears of reed, veer off into the dim green heart of the thicket. Wicken Fen, Cambridgeshire.

purposes of drainage and, in the case of 'lodes' in the Fens, for navigation. Anyone who thinks that ditch-water is dull has only to look at the 'reens' of Somerset, the 'sewers' of Romney marsh, the 'fleets' of the Thames estuary, or the dykes of East Anglia to see that they are teeming with plant and insect life. A survey of Norfolk dykes has found 103 species of aquatic plants, while ditches in Somerset have produced a tally of 135 species, many of them rare.[2] Studies of the Somerset Levels show a natural cycle in ditch vegetation; some plants recover swiftly from dredging, while others gradually increase to a peak population and then decline as the ditch becomes overgrown.[3] Regular management is essential to create the most diverse ditch habitat. An ideal situation is to have the channels all at different stages of recovery from clearance. Ditches in the Monmouthshire Levels, all of which have been subject to a regular dredge since 1970, showed a remarkably stable flora when surveyed in the late 1970s.[4] By looking at old records, it was found that 56 of the 100 plants studied had remained substantially the same since the 1830s. Cattle grazing, together with regular ditch maintenance every 5 to 7 years to prevent major floods, seems to create ideal conditions for such diversity. The real threat to this ditch flora is a conversion to arable farming, which can cause increased chemical runoff, together with the filling-in of many ditches.

## Grassland Management

Grazing, the traditional and, in modern agricultural terms, still the most appropriate agricultural use for many wetlands, creates ideal conditions for wading birds. Snipe, whose long bills end in a sensitive tissue of nerves, will probe for worms in the soft mud of damp grazing land. They could hardly do this in a woodland thicket, which is what pasture would become if totally abandoned. Dutch ornithologists have even found that unmanured hay-meadows, though richer in wild flowers, are actually less attractive for most waders than manured ones. On wetland reserves and artificial bird 'scrapes', where there is insufficient grazing, birds are likely to fly off in search of better grazed farmland. A major farm landowner on the Somerset Levels is now the Royal Society for the Protection of Birds, which supports extensive grazing herds, essential for the survival of redshank and godwit, which breed on the levels. The young chicks would soon get soaked and then chilled if there was no close-cropped grass to walk on, although some

A snipe probes for worms in damp grassland. Some grazing is essential for the conditions it requires.

rough tussocks are needed for shelter and nesting. Redshank thrive on finely structured sward, created by sheep grazing. Ducks too are encouraged by grazing, although their preferences vary.[5] Mallard and gadwall favour lightly grazed fields. Heavier grazing encourages shoveler and pintail, while winter 'lawns' resulting from an autumn cropping bring in flocks of wigeon which, like Bewick's swans, enjoy feeding on marsh foxtail grass, abundant on well-grazed wetlands. Geese too benefit from grazed pasture. At Slimbridge, the New Grounds resulting from past drainage carried out by the Berkeleys beside the Severn are part of the famous reserve managed by the Wildfowl Trust. In 1973 reduced grazing encouraged stands of meadow barley, and so discouraged wintering flocks of white-fronted geese.

The conflict in such places, therefore, is not between farming on the one hand and a wilderness for wildlife on the other. It is a question of the degree of agricultural intensification. Expensive drainage of grassland to achieve maximum stocking on an even and luxurious sward, unflooded in winter, leaves no space for the birdlife. Low-input grazing, which looks increasingly to be in the national interest, when we have, as in 1986, one million tonnes of butter in storage, creates perfect conditions for wildfowl.

### The Ouse Washes

Nowhere do the aims of graziers and conservationists converge more closely than on the Ouse washes in Cambridgeshire. These are the wetlands created by Vermuyden as a safety valve for flooding between his New Bedford and Old Bedford rivers. In summer the land between the raised banks and the great seventeenth-century watercourses is grazed or cut for hay. In winter, especially when the outgoing river water meets an incoming tide, the Ouse washes serve the purpose for which Vermuyden intended them, and fill with the immense volumes of the flood, which flow for 40 miles, all below sea-level, from St Ives to the sea at King's Lynn. With the winter waters, the washes also fill with birds. A staggering 42,500 wigeon, 5,500 pochard, and 7,570 teal have been recorded in one season. In summer, the washlands, two-thirds of which are owned by conservation bodies, are a breeding place for lapwing, redshank, ruff, and godwit. Or, at any rate, they were until recently, for the Ouse washes are now suffering from a problem which does not commonly beset modern wetlands. They are getting too wet. Increased runoff from housing and from drainage schemes for agriculture in the upstream catchment area is sending down more and more water to disrupt breeding birds in the early summer. In 1985 and 1986 all the godwit nests were washed out. The problems for breeding waders are exacerbated by prolific growth of reed sweet-grass, *Glyceria maxima,* for which the sorely needed management is heavy grazing. Botanical studies of the washes published in 1981 showed that on heavily grazed fields the reed sweet-grass was almost non-existent.[6] However, this is hardly true of the Ouse washes in the mid-1980s,

A wetland problem and its solution. The prolific stands of reed sweet-grass in the foreground inhibit the bird-life. Heavy grazing, in this case by sheep, improves their habitat. Ouse Washes, Cambridgeshire.

especially at the downstream end. *Glyceria* is palatable and sweet, as its English name suggests, when the growth is young, but as it grows tall and rank, the beasts avoid it. With a relaxation of grazing, the *Glyceria* bolts away. Obtaining grazing tenants in the great ploughlands of the Cambridgeshire fens, where alternative grazing during the all too frequent periods of summer flood is not to be found, can be problematic. Cutting and burning of *Glyceria* have been adopted to encourage widgeon, and the conservationists responsible for the Ouse washes no longer talk like bird-watchers, but like stock farmers. With an offer by the water authority in 1986 to improve the drainage on the reserve for the benefit of conservation, the wheel appears to have come full circle.

## Pollard Willows

Perhaps the most elegant and universal example of the art of active wetland management concerns the pollard willow. A pollard is a tree of any kind which has been polled, or evenly cut, to form a crown about 8 feet above ground level, just beyond the reach of grazing animals. This resourceful 'cut and come again' approach to timber management is probably as old as civilization itself. At the 5,000-year-old riverside camp at Etton, archaeologists have found the remains of pollard willow and alder, complete with recognizable cut-marks left by bronze axes. On the Somerset Levels, remains of pollards have been discovered dating to as early as 4000 BC.[7] In parts of Europe, and even more so in the Third World, pollarding is still common practice as an essential way of obtaining both firewood and fodder. Even in England, signs of past pollarding can be seen on old oaks and ash, which were often lopped for timber. Nowadays swift-growing willows are the only trees to be commonly pollarded. They were always popular. In the churchwarden's accounts for Buckden, near Huntingdon, a non-tenant who lopped an ash tree in 1629 was fined tuppence, while whoever damaged the useful willow had to pay eight shillings a tree.[8]

John Clare's writing is full of pollards. He called them 'dotterels'; and in his parish, as in the rest of the country, they clearly had plenty of uses. Clare writes of a shepherd 'hiding from the thunder shower in an hollow dotterel' and of 'a gate whose posts are two old dotterel trees', and in his poem 'The Fens' he describes:

> The trees to stumpy dotterels lopt
> The hearth with fuel to supply
> For rest to smoke and chatter by.[9]

The by-products of all this practical exploitation have been trees of great age and even greater character. Clare's 'leaning dotterels' were green with ivy in midwinter, and pollards, as a direct result of the specialized management which they have received, are extraordinarily rich habitats for wildlife. Foxes sometimes make their homes in them, as Beatrix Potter observed when describing the sinister Mr Tod, who 'moved into a pollard willow near the lake frightening the wild

Pollard willows, one of the most characteristic features of the wetland landscape, rely for their diversity of wildlife upon active management. Ashleworth Hams, Gloucestershire.

ducks and the water rats'. Despite such dangerous visitors, mallard often nest in old pollards, as does the little owl, which glowers with unblinking yellow eyes through a crack in the willow bole and from which, at dusk, it glides like a silent, oversize moth. Anyone who has tried to do much pollarding will bear witness to the frequency of wild bees' nests, while moths associated with willow include the eyed hawk-moth, the lunar hornet, the clouded border, and the July high-flier. Best of all is the puss-moth with its exotic-looking caterpillar, whose two-pronged tail blushes pink when agitated. Willow leaves are often blistered with vermilion bean galls, caused by the eggs of sawflies, or swollen with the crusty growths of the willow rose gall, which hang like old birds' nests among the willow twigs after the leaf has fallen, providing a home for the gall-midge and prey for other species.

But the most remarkable thing about willow pollards is the way in which they are colonized by other plants. An epiphyte is a plant which grows on another plant without deriving nourishment from it, sustained only by soil and moisture accumulated in crevices in the trunk and branches. Epiphytes are a major constituent of the extraordinary vegetation of tropical rain forests, where they create whole hanging-plant communities, festooning every branch with their green luxuriance. This seems a far cry from the pollards of an English riverside, and yet, between 1890 and 1892, two young scientists, Willis and Burkhill, both destined to become leaders in tropical botany, spent what must have been some pleasant summers walking beside the Ouse and the Cam in Cambridgeshire. They looked at some 4,000 pollard willows, on which they found no less than eighty different species of plants which could strictly be described as epiphytes.[10] These were all typical flowers of the English countryside, many of which you can still find in the unlikely eminence of a pollard top if you search hard enough: cow-parsley, goose-grass, nettle, rose, gooseberry, and wild currant. The continuity of occupation by the pollard flora surprised them. One tree still supported a crop of the annual, wall lettuce, exactly as an earlier botanist had recorded it 35 years before. Given the longevity of well-polled willows, this is not exceptional; mankind has been setting up the right conditions for such floristic diversity by active cropping for many generations. Oliver Rackham writes that the practice of growing pollard willows in Cambridgeshire was first recorded in the thirteenth century.[11] In attempting to establish how so many plants without obvious means of seed dispersal had rooted in the willow crowns, Willis and Burkhill undertook an ingenious analysis. They botanized the birds' nests! Identifying every leaf and wisp of grass in the nests of wrens, thrushes, and sparrows, they discovered wild carrot, thistle heads, and even some living strands of pondweed.

You can be sure that the owners of those willow trees by the Cam would have taken a more than academic interest in them. Their regular cropping on a 7–10-year cycle would have provided firewood, which does not live up to its ominous reputation for splitting if allowed to dry out for a couple of years, and also

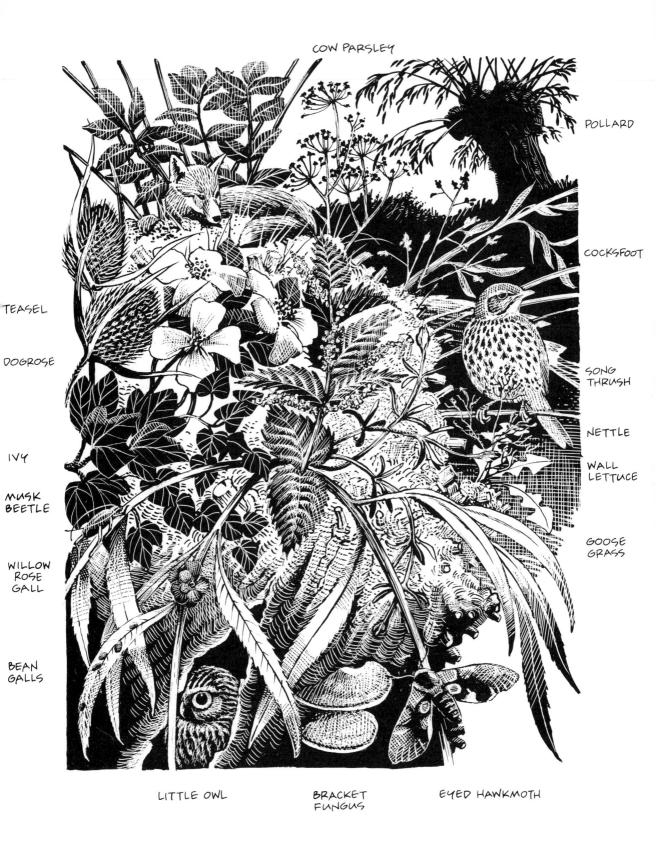

COW PARSLEY

POLLARD

COCKSFOOT

TEASEL

DOGROSE

SONG
THRUSH

NETTLE

WALL
LETTUCE

IVY

MUSK
BEETLE

GOOSE
GRASS

WILLOW
ROSE
GALL

BEAN
GALLS

LITTLE OWL

BRACKET
FUNGUS

EYED HAWKMOTH

The remarkable range of wildlife to be found in a pollard willow.

produces some of the hottest ash of any wood, ideal in a bread oven. In addition, they provided useful fence rails. As fence posts, willow poles would be useful only at occasional intervals, given their irrepressible tendency to grow whenever hammered into the ground. Of such things is the rightness and the real quality of the landscape made, a resource in all senses of the word, which it is time we set out to rediscover. 'When you think of all the willows which we've felled and buried!' exclaimed a Leicestershire drainage man in 1986 as, for the first time in his career, he hammered in some willow stakes taken from the last surviving tree on a particular river bank. There is no reason why the progeny of that willow, and their progeny in turn, should not provide plenty of practical uses for the riverside landowner, as well as contributing to the beauty of the river bank.

For a start, wood-burning stoves will take willow wood, and so a line of pollards with their regular crop of poles will keep a landowner in fuel, even if he does not sell them to other stove-owners. A more unexpected, but increasing, outlet for pollard poles is the manufacture of cardboard boxes. Most paper and cardboard is made from softwood, such as Sitka spruce, but for the corrugated sections of a box, a hardwood pulp with a short fibre length is essential. For this, timber agents for the big paper-mills travel around the country, and in 1986 were offering £13.50 per tonne for willow, provided the farmer left it sawn and stacked for collection. For twenty to thirty pollards—the minimum number which they are generally prepared to crop—they can be offering £400: not a fortune, but a lot better than nothing, especially with a promise to return and crop the same trees in another ten years. There is, of course, the risk that those in search of willow poles will also take other broad-leaved trees, though the requirement for nothing larger than 16-inch-diameter timber precludes the felling of fine specimens. Another problem is that all timber is bought by the tonne, and willow poles, which may look large in volume, dry surprisingly fast to what can be a disappointingly low tonnage. None the less, timber merchants predict that there will always be a demand for pollard poles, and they are developing an on-site chipping machine, which will harvest a whole crop of poles and chip them on the river bank. With this arrangement, the farmer gets less money for his timber, but at least his trees receive a new lease of life. With such developments, there is hope that the old cycle of decline, whereby unpolled willows split and are finally felled, may be reversed, and that a new generation of river managers will set productive willow stakes along our many miles of treeless river bank.

## Osier Beds

Pollarding is not the only way in which willows can serve man with a rich harvest. Within living memory, working osier beds, like the one drawn by Shepard in his river map in *The Wind in the Willows*, were an essential part of every river landscape. These damp woods, known as 'garths' in the north and the east, and elsewhere as 'withy-beds' or 'willow holts', were intensively managed to produce willow rods for basket making. Willows, especially the osier *Salix viminalis,* were grown close together, and were cut down to the ground every winter, yielding a seasonal stand of straight, slender, man-high willow wands.

Attenborough withy-bed beside the Trent, where willows have been grown since the days of the 'Domesday Book', is now managed by the Nottinghamshire Trust for Nature Conservation. Such is the wild intricacy of this 250-acre willow wilderness that a researcher returning there, having previously spent three years studying its birdlife, managed to lose himself in it. But Attenborough, like Wicken, is not as wild as it feels. On a still day you can hear 'Little John' chiming in Nottingham market square, four miles away. In its present form it has been largely created since 1966 by the trust which manipulates the water-levels and cuts both reed and willow on a regular rotation, selling these commodities to thatchers and as heatherings for hedge laying. Studies at Attenborough have shown that the resulting managed stands of cream reed and green osier support the highest con- centration of nesting reed-warblers and feeding sedge-warblers and reed-buntings of all English habitats.[12] Some warblers, notably chiff-chaff, blackcap, and willow-warbler, even winter there. The reserve abounds with stockdoves, tawny owls, several pairs of kingfishers, at least eighteen but- terfly species, and a dozen different damselflies. The crusty bark of the older willows is pitted with holes made by lunar clear wing and goat moth larvae, which, together with musk beetles, are eagerly sought by both greater and lesser spotted wood- peckers.

Lesser spotted woodpecker.

Yet the enduring impression of Attenborough is the way in which it is, and has been, loved: by the cyclers and walkers who frequent its maze of paths, by the men who manage it now, and by past gen- erations of withy managers who have left their mark on the place. In the mid-nineteenth century, the withies were worked by William Scaling, the renowned Victorian basket- maker, and it was here that he developed his own variety of willow, which is still the glory of Attenborough: *Salix basfordiana,* named after the nearby village of Basford, where he had his home. His withy-bed is a kind of willow orchard, with that special blend of garden and woodland, domesticity and wildness, which orchards have. Willows, like apple trees, come in numerous varieties, developed over many years and for different uses in the basket trade: Glibskins, Golden, Spaniard, Blob, Blackrod, and Whissender. Attenborough is magnificent in August, when loosestrife, richly purple as summer pudding, blooms beside the tussocks of osier. But perhaps the best time for a visit is March, when the duckling down of pussy willow studs every tapering twig, massed on the waterside sallows and daintily etched along the tips of the floundering stems of Purple Willow. Then the wind does not whisper in the willows: it sighs and snaps at the brittle branches, and the creak of the Crack Willow, groaning like a whole navy of ships' timbers, drowns all the other sounds of the withy-bed. With the rising sap, the various withies intensify in colour: the lemon osiers, the plum colour of the Violet Willows, and above them all the enormous orange aureole of the big Basford Willow trees, glowing against the slate spring sky. A few weeks later, the pussy willows lengthen, and the whole withy-bed breaks out in the tenderest green.

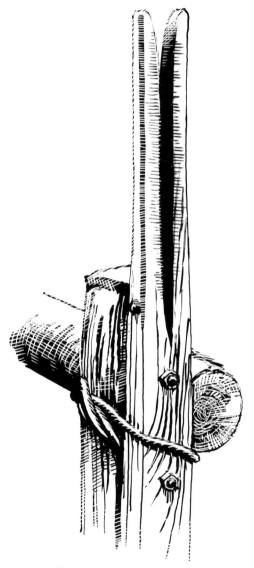

Willow brake, used for stripping willow.

For years, I have been defending small riverside willow woods from the ravages of drainage schemes, without fully realizing what they were: the last visible remains of a highly evolved craft, in which a local culture and a sense of place were refined about as far as is possible. The weaving of willow is as old as civilization. Basketwork has been discovered in the Iron Age lake village at Glastonbury[13] and in the tombs of the Pharaohs. Early pottery imitated the patterns of the woven willow: a testament to man's enduring conservatism, as is the modern plastic laundry basket, which mimics a basket weave. In more recent times, willow rods were stripped by drawing them through a cleft metal device called a hand-brake, and the lunging motion of this action became the model for the country dance 'Strip the Willow'. Willow stripping for white rods was possible only between April and June, when the green withies were soft. Women would gather with their children to do this work, which also had the status of a seasonal social event, like fruit picking or hop picking. With the development of boiling techniques, rod stripping for buff willow became possible throughout the year; and so by the nineteenth and early twentieth centuries, riverside osier beds, estimated at a total of 6,000 acres in 1925, supported a labour-intensive industry. In 1926 the Agricultural Economics Research Institute produced a kind of 'Domesday Book' of current rural industry, which included a survey of the basket trade.[14] It reads like a litany of the regional variety of an England we have lost. In the East Midlands, stout willow was woven into potato hampers. In Kent and around the Vale of

A corner of an old withy bed. In its decline this looks very different from the serried ranks of willow rods which would have dominated the working withy bed.

Evesham, willow was grown for fruit baskets, and even now, in the Vale, withies neatly tie bundles of hundreds of 'Evesham grass' at the Bretforton asparagus show. At Berkhamstead they were used for watercress baskets. Ipswich withies went for cockle baskets. Lincolnshire fenmen made eel traps and 'fish kiddles'— hence one explanation of the expression 'a pretty kettle of fish'. Cornishmen fashioned lobster-pots out of the heaviest willow, as they still do in a few places. In Lancashire, the finest willow, of which it was said that 'you could lace your boots with it', was made into baby cots and pigeon baskets. These latter were lined with down, and had the most intricate hinges and catches. One Somerset expert still makes them for export to Japan.

In some areas, growers and basket-makers combined forces. Elsewhere they diverged, and the 1926 survey remarks of one north countryman that he 'seems almost to be a Trent valley man gone astray for he is an artist in osier growing and makes baskets for a penance'. How swiftly vanished is that great tradition. In the nineteenth century the area of the Trent and its tributaries was the headquarters of the English osier industry, and it was said of the growers there that they 'produce some of the finest osiers in the world' and 'have little to fear from foreign competition'. Now, apart from Attenborough, which is managed as a nature reserve, scarcely a willow holt survives there. A short-lived demand for baskets in which to drop supplies over the enemy line in the War did little in the long run to halt the inexorable felling and bulldozing of willow trees to make way for wheat. Our surviving basket industry centres on Somerset, and is only now beginning to see a real revival.

But behind the withy-beds, as behind all the elements of the English landscape, stand men: in this case, men who love willows. The names of many varieties of withy commemorate their creators. The most famous is Dicky Meadowes, a variety of *Salix purpurea* which Richard Meadowes of Mawdesley in Lancashire propagated from ties he found fastened around a bundle of raspberries which he was importing from abroad. Now, as a question hangs over the future of the agricultural crops for which the withies were so ruthlessly cleared, a new generation is starting to look hard at osier beds, not only as one of the glories of the riverside landscape, but also as an economic resource. Two men, who bring very different perspectives to the matter, are a West Country craftsman, David Drew, and a Nottinghamshire farmer, Chris Mollart.

Visiting David Drew, it is possible to see how the quality of our lives has been diminished, not only by the loss of willows, but also by the loss of the baskets which came from them. A hamper is the only civilized thing to use for a riverside picnic, as Ratty and Mole well knew. Now we make do with a plastic bag and with river banks which are so degraded that they are not worth picnicking beside. In David Drew's cottage in Somerset is a Polish basket made for gathering mushrooms. It is an object of extraordinary beauty, not only because its shape is moulded entirely by the materials of which it is made, but also because all those materials are a product of the ecology of the Polish landscape out of which

David Drew making a basket.

they came. The spokes are made from branches of juniper, through which are woven the supple roots of pine. David's own baskets share this quality of variety within unity, which is also a basic principle for any decently managed, and so aesthetically satisfying, landscape. He exploits every kind of osier he can obtain, many of which he grows himself, and uses them both stripped and unstripped. With them he creates woven patterns of purple-green, chocolate, olive, and the white of stripped willow, which grows old gracefully, developing a creamy gold patina with use and time. In 1986 the Crafts Council mounted a travelling exhibition for these baskets, and they are now eagerly sought after as works of art.

It was not always so. Since the early 1970s, when David Drew started making log baskets, for which a market was created by wood-burning stoves, consuming the glut of dead elm logs, he has fought a long and lonely battle to make the craft of basket making respected in artistic circles. The very word 'wickerwork' conjures up occupational therapy or the dreariest of primary school penances. Ever since the Napoleonic wars, basket making has provided a subsistence income for the wounded, just as willow wood has yielded a sad harvest in the aftermath of war for makers of artificial limbs. The price of baskets made by sighted craftsmen has been artificially suppressed until recently by the standard rates demanded by Workshops for the Blind.

David Drew combines idealism with the toughness of a good businessman. All his output for 1985 was sold to the Crafts Council in advance of its exhibition, and the prices per item ranged from £7.50 to £142.50, with plenty of baskets at the upper end of that range.[15] Without recourse to advertising or even seeking retail outlets, he is confident of a demand for as many as he can make. Of course, not everyone can be a master craftsman, but there must be room for a few more like David Drew, and that has to be good news for the osier beds down on the river bank. David insists that one of his main goals has been to encourage others by demonstrating what can be achieved. A last glimpse of the artistic and commercial possibilities of withies is the elegant gate opening into David's vegetable garden, knocked up in a couple of hours, he says, out of hazel and willow wands, and not something which he is especially interested in repeating. But somebody else might be. Willow garden furniture was all the rage at the Chelsea Flower Show in 1986.

Chris Mollart's approach to withies is not that of a basket-maker, but of a farmer looking for a crop to supplement his 186 acres of cereals in Nottinghamshire. His inspiration was a talk given by a local historian about the once glorious Trent osier industry, and his withy-bed is a lesson in the possibilities of combining landscape quality with practical farming. Chris points out that growing vast acreages of willow may expose them to pests and diseases. A monoculture of withies is as impractical in farming terms as it would be unattractive as a wholesale alternative to landscapes now given over entirely to wheat. Neat hedges divide his impeccably grown arable acres, and in late July the pea-green wheat, the jade oats, and the tan of the ripening barley all converge on his lush willow holt, nestling snugly in the valley. It was planted in the spring of 1985, thereby saving the little hedged paddocks from being ploughed for further corn, and the following year Chris Mollart had more orders than he could supply.

His withies go for less exalted baskets than those crafted by David Drew, but there is a steady and increasing demand for them. To meet existing requirements, Chris estimates that there is room for another 100 acres of osier beds in the United Kingdom; but to replace all the baskets now imported from Romania, Madeira, and the Far East would take a further 2,500 acres.

David Drew's comment on imports is that they have always been a problem, and always will be. People will not pay to eat chicken out of a basket which is a work of art. He shrewdly points out, however, that the Chinese are exporting imitations of traditional English baskets. David's special contribution has been to show that we can be proud to produce the real thing.

Chris Mollart's solution to imports is quite different: mechanization. For £2,000 he is making his own harvester, partly adapting a flax-puller and an old binder, basing it on machines used in Belgium, where most European withies are grown for export as vine ties. This should reduce his annual cash input per annum per acre to £200, which is low for any crop nowadays. With machinery to cut and then sort the rods, he says, we could bring the cost of the product down, thereby competing efficiently with cheap labour abroad. An osier crop must be as clean as a field of wheat. Chris Mollart sprays out the weeds in his withy-bed, but he may adopt David Drew's method of using black polythene mulched with hay from an adjacent flowery meadow. There is still room for flowers around the edges of a willow plot; indeed the hand-cleaned ditches intersecting all the Somerset withy-beds are among the most ecologically rich in the Somerset Levels. What both men's osier beds have in common is their real contribution to the landscape. Chris Mollart's willows are protected from adjacent cereal spray by ancient hedges of oak, hawthorn, hedge maple, and white bryony, the very same species which were carved with such precision and love by thirteenth-century craftsmen in the stonework of nearby Southwell Minster. For an annual turnover estimated by Chris at £2,000–£3,000 gross per annum per acre (compare wheat at £200–£300), the words of an anonymous Fen poet of the 1680s have a distinctly modern ring:

> The Gentle Ozier, plac't in goodly ranks
> At small Expense, upon the comely banks,
> Shoots forth to admiration here and yields
> Revenues certain as the rents of fields.[16]

### Cricket-bat Willows

Elsewhere in England, farmers are growing willows for yet another purpose, which is both agreeable and utterly English: the manufacture of cricket bats. The river Leadon, beside which Edward Thomas walked and wrote his poetry on the eve of the Great War, has its source among the red hills of Herefordshire. Here Richard Porter farms 150 acres of tipsily rolling valley slope with well-groomed cider orchards, sheep pasture, black currants, and wheat. Threaded like a silver seam along every brookside in his valley is a procession of cricket-bat willow trees. Fitting John Clare's description of 'the grey willow shining chilly in the sun as if the morning mist still lingered in its cool green', *Salix alba coerulea* has the most silver-blue shimmer of any willow. It is the only species which makes good bats—Kashmiri willow is heavy and breaks on contact with a fast ball— and England is the only place where it is grown. We already export 70 per cent of our production to Australia, and one Essex dealer anticipates a problem of insufficient bat willow in the near future, not helped by the activities of drainage men in his own county. Others are pursuing the possibilities which arise from the latest boom in baseball in the United Kingdom.

Richard Porter has seen 400 bat willows pass through his hands, although he once had ambitions to grow 1,000 trees, as a civilized alternative investment to subsidizing the questionable virtues of life insurance salesmen. The turnover time of the crop is a minimum of fifteen years, and a grove of willows he planted beside the Leadon on Christmas Eve 1975 had reached a stately 30 feet ten years later. They thrive best near running water. With no staking required and just the occasional trim 8 feet above ground level, the cost of these trees works out at about £2 each, and the return at 1986 prices is around £55 per tree, with the additional reward of firewood from the unusable top half. The Suffolk bat merchant who comes and harvests Richard Porter's willows when they are ripe for felling still operates a neat arrangement with those farmers who grow sufficient for his purposes (ideally a hundred trees). With the cheque for payment for the latest crop, he despatches a jiffy-bag containing cuttings of *Salix alba coerulea*. The nursery then is simply the river bank. What could be more elegant?

## Riverside Woodland

In addition to his cricket-bat willows, Richard Porter planted a small wood of ash and wild cherry in 1986, with a grant from the Forestry Commission Broad Leaves Scheme. This arrangement, launched in 1985, when the Forestry Commission for the first time offered subsidies to plant native deciduous species for timber, was a victory not only for nature conservation, but also for common sense. For decades many foresters, just like the more old-fashioned drainage

Some uses for valley-bottom land which pay their way. Fishing pool and cricket-bat willows. River Leadon, Hereford and Worcester.

men, stuck to the methods and solutions which they had been taught; and it was regarded as heresy to suggest that oak and ash, which had evolved over the millennia to dominate all other plants in an English wood, could actually yield as efficient and commercial a crop as exotic conifers. It is difficult to avoid the conclusion that foresters' prejudice in favour of exotic spruce has something in common with the attitudes of some agriculturalists and drainage men towards willow trees, that anything wild must be a weed.

In the mid-1980s it is interesting to watch some water authority estate staff engaging in a yearly battle with man-high ash saplings, between which stunted Christmas trees inch their way towards the light. It is not hard to see which is the most efficient converter of energy into timber. By contrast, the French, who have never turned their backs on our common tradition of harvesting the resources of the ancient wild wood, have developed the management of their deciduous woods into a modern technological skill. In the riverside woods of most French valleys, on carpets of wood anemones and bluebells, you will find neat stacks of carefully graded logs. As for wetlands, resourceful management of such woodlands improves their habitat value into the bargain. An equivalent plantation in lowland England is likely to be shadowed by the sterile gloom of spruce, fir, or pine, which, apart from their use as a fringe to hardwood game

Resourceful use of riverside timber. In the spring of 1987, the government announced that hardwoods may offer an alternative use for agricultural land which is surplus to requirements.

coverts, offer little for the wildlife and nothing for the landscape. The turnover of any forestry harvest, even on a 15–20-year rotation, is never going to match the annual return from grass or grain of course and for this reason a small annual grant to offset maintenance costs is advocated by some timber-growers. However, if and when some riverside land goes out of conventional agricultural production, let us hope that those who adopt the timber option will have the wisdom to plant only our beautiful, fast-growing native hardwoods. It is also important that, when trees are planted, it is on land set aside from the previous production of surplus corn, and not, as is already happening in some cases, in the last good corners of surviving habitat on the farm.

Even before the tide began to turn with the Broad Leaves Scheme, the Country-side Commission in 1983 produced a report based on a nation-wide survey called 'Small Woods in Farms'.[17] This found that only 1 per cent of the farmers and landowners interviewed had sufficient basic knowledge to manage their own existing woods effectively. The report suggested that the standing value of timber in farm copses was around £450 million, a figure which could be increased considerably with proper management. Alder, no longer needed for clogs, is still widely used abroad, and its potential as a timber was felt to be well worth exploring. A major use for timber is as pulp for paper making. For cheap paper, such as newsprint, Sitka spruce is preferred because of its long fibres. However, chemical pulps, which are a major component of high-quality paper, can be made from fast-growing hardwoods such as alder and aspen, *Populus tremula*. This lovely native tree is very much at home on damp riversides. A whispering grove of aspens guards one of the droves at Wicken Fen. The Common Market is currently funding research into ways in which paper-manufacturers can reduce dependence on imported chemical pulps, by using native species which may one day create attractive woodlands on middle-grade riverside land which presently supports surplus corn or silage.[18]

For farm forestry to break into large-scale timber growing, change in the way in which private forestry is subsidized has long been necessary. In March 1988 the tax incentives which had led to massive afforestation of Scotland's Flow Country were at last reformed. Carefully planned grant aid and liaison with conservationists will now be needed to ensure that our new lowland woods are ecologically appropriate.

## Willows as an Energy Source

The ever-serviceable willow may one day provide the ultimate economic jus-tification for large riverside woodlands. Since 1975, trials carried out in Sweden, Canada, England, and Northern Ireland have shown how stands of regularly coppiced willow can be used as an efficient fuel source, either fed directly into heating boilers or compressed into fuel briquettes for use in the same way as coal.[19] The willow which is currently considered the most appropriate for this purpose is a vigorous Baltic species, *Salix burjatica* (syn: *Salix 'Aquatica*

*Gigantea*'). Such is the vigour of these trees that the Irish trials have shown a mean yield of 16 tonnes of dry matter per hectare per annum on a 4-year rotation. In addition, researchers have found ways of converting this willow timber into ethanol for use as an octane-enhancer to replace lead in petrol, and as an animal feed supplement. Problems remain in the short term. Because we are wedded to a conventional energy policy and still have reserves of oil and coal, there is little political interest in bio-fuels. In addition, *Salix burjatica* is susceptible to rust diseases. Of course, there is a vicious circle, whereby lack of interest in these alternatives discourages programmes of research and development. The time is fast arriving, however, when a closer investigation of alternative land uses will be more of a political and an economic possibility.

One of the great functions of nature conservation is not so much the preservation of things as the preservation of options. Within any complex habitat, including wetlands, there are plenty of 'weeds' which offer mankind a variety of uses, virtually impossible to predict until the moment arrives when man finds himself suddenly in need of them. Drainage which destroys, or else reduces, wetlands to such small units that their species populations become genetically unstable may eliminate a resource which we never suspected we required until it was too late.

### Alder Buckthorn

This was illustrated dramatically by an unforeseen consequence of drainage activities in the last war. Alder buckthorn, *Frangula alnus,* is a wetland shrub whose chief claim to fame among naturalists is as a food plant for the brimstone butterfly. It has also long been known to the manufacturers of gunpowder as the plant which, even more than alder and willow, produces the best and most evenly burning charcoal. In the middle of the War, when drainage was at its peak for the production of food, it was suddenly realized that the usual sources of this shrub, now desperately needed for slow-burning fuses in the manufacture of munitions, were under enemy control. In 1941 Wicken Fen was scoured for

Alder buckthorn and brimstone butterfly.

buckthorn for making bombs—scarcely an example of a wetland product contributing to the well-being of society—but it set some of the war agricultural committees back on their heels as the Fens and the New Forest were searched for what by then, as a result of drainage, had become a scarce plant.[20]

## Aspirin and Other Drugs

Other wetland plants have more constructive uses. Meadowsweet has always been loved for its honey fragrance, and Chaucer describes its use in mead. It was strewn on the floors in the court of Queen Elizabeth I, for, as Gerard writes, 'the smell thereof makes the heart merrie and joyful and delighteth the senses'.[21] In 1899 salicylic acid, obtained from meadowsweet, whose old Latin name was *Spiraea ulmaria,* was compounded with other drugs to make aspirin—hence its name 'a spiraea'.[22] Salicin, the active ingredient of willow, was also used in aspirin, but now, since it can be produced synthetically, this particular source is obsolete. This does not mean that there may not be further medicinal uses for these and many other wetland plants. Tincture of valerian made from the roots of *Valeriana officinalis,* whose shell-pink flowers adorn damp meadows and ditches in July (not to be confused with that cheerful flower of West Country walls, red valerian), has long retained a respectable place in the British pharmacopoeia as a sedative. It is still contained in many brands of proprietary medicine.[23]

## The Medicinal Leech

There is another inhabitant of English wetlands for which modern surgery is finding unexpected and critical uses: the medicinal leech, *Hirudo medicinalis.* This is one of sixteen leeches native to the United Kingdom, and it has been used for blood-letting since classical times. 'It were too tedious', lamented the seventeenth-century Doctor Mouffet, 'to reckon up all the melancholique and mad people that have been cured by applying leeches to the hemerrods in their fundaments.'[24] By the nineteenth century, when their vogue among doctors was at its peak, leeches were becoming scarce. One reason must have been over-collecting, although the bloated specimens were frequently returned to ponds and ditches. Another cause must have been land drainage, which was extremely common, especially in the early years of the century. William Wordsworth, in his famous poem 'Resolution and Independence', written in 1802, described the leech-gatherer as an old man who appeared to be fading away as fast as the leeches for which he was searching:

> Gathering leeches, far and wide
> He travelled; stirring thus about his feet
> The waters of the pool where they abide.
> 'Once I could meet with them on every side;
> But they have dwindled long by slow decay;
> Yet still I persevere, and find them where I may.'

Traditional haymeadows have been one of the casualties of drainage and high-gear farming. *Top.* A rich mix of flowers and grasses just before haymaking. Oxfordshire. *Bottom.* Snake's-head fritillaries around a traditional stone hay-marker. North Meadow, Cricklade.

Two approaches to farmland in the West Midlands. Both landscapes are Grade 2/ Grade 3 agricultural land. *Top*. Fruit, sheep, and some corn make up the diverse landscape of Richard Porter's farm near Malvern. *Bottom*. A square mile of clods. Orton on the Hill, Leicestershire.

*Valeriana officinalis.*

Valerian, illustrated here in *Medical Botany* by Stephenson and Churchill, published in 1836. It is still used as a sedative.

An especially gruesome way of gathering leeches. Gatherers wade into the water, attracting the leeches to their bare legs. From *The Costume of Yorkshire*, 1814. Photo: Wellcome Institute.

Wordsworth's leech-gatherer splashed the pool with a stick, and this effective method is still used by some modern biologists when searching for medicinal leeches. The hungry animals wriggle up through the water towards the disturbance. A rather more robust approach to leech gathering is portrayed in George Walker's *The Costume of Yorkshire,* published in 1814, which illustrates women wading bare-legged into suitable ponds to attract them. In our time, medicinal leeches have continued to dwindle, largely because ponds and ditches, along with the leeches' staple food of young frogs, have been drained away.[25] Though declared extinct in this country in 1910, the medicinal leech survives in a few isolated ponds in Romney Marsh, the New Forest, Wales, and the Lake District. However, a biologist searching for leeches in 1985 found them absent from their previously regular sites in Hampshire, and even from the Normandy marshes of La Sangsurière, a region so renowned for leeches that it took its name from *la sangsue*, 'the leech'.[26]

Leeches are used in modern medicine and research as a source of the anti-coagulant hirudin, as well as other potentially useful biochemicals. They are also increasingly used for controlling blood flow during micro-surgery and skin-grafting operations.[27] A firm in Slough imports 5,000 a year for medical purposes, and many of the leeches used in English hospitals are collected in European wetlands. It is possible, but not always very easy, to breed *Hirudo medicinalis* in the laboratory.

*Drainage in the Third World*

So much for the unforeseen potential of the English wetlands. In the Third World, where untapped drug resources are almost certainly being destroyed before we have even discovered what they are, the case against wholesale drainage is even more compelling.[28] The United Kingdom, which helps lead the world through such organizations as the World Wildlife Fund, also exports engineering expertise to carry out wetland drainage which we would certainly not countenance at home. One English firm of consulting engineers proudly announces in its latest brochure the reclamation of 10,000 hectares of Wahgi Valley swamp in Papua, New Guinea. Another leading English consultancy estimated in 1986 that 70 per cent of its land-drainage work was in the Third World. This is against a background of an estimated 50 per cent loss of wetlands worldwide since 1900 and such shocking examples of land mismanagement as the drainage operations in the Orinoco delta in Venezuela in the late 1970s. Here drainage triggered the formation of acid sulphate soils, which ensured that the local Guaraunous Indians could no longer grow bananas or cassava. Nor could they depend any longer on wild animals as their chief source of meat, since the wildlife was devastated. The Indians were moved to another area, where a series of floods in 1976–80 claimed the majority of the population of 4,000–5,000, and the few survivors are now classified as displaced persons.[29] The shades of Vermuyden and the commoners beckon from the past.

It has been estimated that some wetlands are capable of producing eight times as much plant matter as a field of wheat. Aquaculture—for example, farming rice and crayfish in the same ponds in the southern United States—is extremely productive, but the potential from harnessing natural wetland systems can be even greater. The average yearly animal protein production in swamps and marshes is 3.5 times the average for natural terrestrial ecosystems.[30] Various native plants of tropical wetlands, such as the sago palm in the South Seas, are farmed directly from the swamp.

*Economic Uses for Reed-beds*

Wetlands in the tropics and subtropics can act like gigantic septic tanks, absorbing nutrients and toxins. Florida cypress swamps have been used as natural tertiary treatment systems for domestic waste waters, and cost 60 per cent less than treatment plants which must be constructed. Common reed, *Phragmites australis*, has been used in South Carolina to treat slurry, and this same plant is found in abundance in English wetlands.

Many English reed-beds have been drained, and engineers and farmers generally regard the reed as a typical wetland weed. Now, through specific management techniques, engineers in English water authorities are beginning to use stands of common reed as a practical alternative to sewage-treatment beds. These

Thatch made with reed and capped along the ridge with saw sedge.

methods, pioneered in Germany and Denmark, use the extensive root system of the reeds and their associated bacteria to break down the sewage.

Reeds also yield a more immediate harvest, in a way which further reinforces the character of the English countryside, in the form of thatch. With its quiet colour, complementing the varying qualities of brick, stone, and plaster, well-crafted thatch can be crisply sheared so that it slices up the patterns of light and shade, while its texture is as velvet as the coat of a well-groomed cat. Developed by modern techniques to be relatively fire-resistant, as well as being cool in summer, warm in winter, and almost impervious to noise, thatch is enjoying a revival. Many buildings listed in the 1960s are now in need of repair, and some modern builders are even adopting this most ancient of roofing materials. One Dorset thatcher has even discovered a booming trade in Merseyside. Vernacular buildings in hill country were traditionally designed to accommodate straw thatch, into whose humpy cushions chimney-stacks nestle, rather than stand proud, and in such circumstances reed thatch is inappropriate. Long straw of such varieties as 'Huntsman' and 'Wigeon' is now specifically grown for these purposes. Elsewhere in the lowlands, and especially in East Anglia, reed makes ideal long-lasting thatch, and there is a case for returning thatch to many roofs which originally supported it. Reed can also be made into attractive screens and fence panels, and is also used in the manufacture of cellulose.

A revival in thatching benefits not only the architectural environment. A well-

Common reed, increasingly scarce because of drainage. We import most of our reed for thatching from eastern Europe.

managed reed-bed is a paradise for birds, notably such scarce species as Savi's warbler, marsh-harrier, bittern, and bearded tit. The last two are among the most glamorous of our wild birds, and appear quite capable of colonizing the new reed-beds which might be provided for them. Tiger-striped to camouflage him perfectly among the reeds, the male bittern reveals his presence by his booming call in spring, which has earned him some entertaining local names: buttleebump, bogle, bitter-bum, and bog-bumper! As a result of persecution and drainage, silence reigned over the bittern's old territories during the last 30 years of the nineteenth century, but in 1900 its booming was heard again, and in 1978 we had an estimated fifty breeding pairs.[31] A high proportion of these birds were breeding at Minsmere in Suffolk and Leighton Moss in Lancashire, both reed-beds which have been entirely created this century, following the abandonment of drainage. The bitterns' ability to colonize was illustrated by the way in which a pair nested in the year immediately following the diversion of freshwater into a coastal reed-bed at Titchwell in Norfolk. In hard winters, bitterns range far and wide—startling winter records for the late 1970s include a bus-stop in Stoke Newington and a shop window in Gravesend—and this may lead them to colonize suitable reed-beds which have plenty of ditches, over which they can crouch to spear eels and other fish. Maintenance of drainage ditches is as essential

for the bitterns as it is for commercial reed-growers, who need to control water-levels to minimize frost damage in spring, as well as to gain access for harvesting with converted rice-cutters.

The bearded tit, with its elegant blue head and long black moustachios, has settled into new reed-beds with even greater ease than the bittern. Breeding even along the Thames valley in the nineteenth century, our bearded tit population was reduced, in part through drainage, to a mere four pairs in 1947. By the mid-1970s, boosted by birds migrating from the vast reed-beds created to reclaim the Dutch polders, our reed-beds supported some 600 pairs; and since colonizing Leighton Moss in 1973, they have doubled their population on that site alone with every successive year. Not being territorial, they nest close together in the uncut reed and then forage widely over the harvested areas.[32] The Norfolk Reed Growers Association has described how, in 1968, flocks of bearded tits searched the stubble for food almost under the feet of the reed-cutters.

In 1982 a study produced by the Royal Society for the Protection of Birds concluded that there was no evidence that reed cutting had an adverse effect on the presence of the scarcer birds of reed-beds.[33] Indeed, if reed is not managed, it will develop into scrub. Cut areas within a reed-bed also create temporary open water, which benefits ducks and waders. One problem is the thatchers' requirement for 1-year-old reed (known as Single Wale), or at most 2-year-old reed (Double Wale), which leaves less room in the reed-bed for old undisturbed stands. A simple solution adopted at Wicken is to allow regular cutting on only part of the reed-bed. In addition, reed-growers believe that total clearance every year weakens the bed; therefore, many owners settle for a programme of Double Wale, which means that half the marsh is cut each year and always by April which prevents disturbance to nesting birds. The Royal Society for the Protection of Birds cites reed-beds created for the thatching trade in the Norfolk Broads which now support populations of all the specialized rare birds of reed swamp.

Bittern.

Such are the all-round benefits of this wetland resource which in 1986, without any external subsidy, was estimated to be turning over between £60 and £186 net profit per acre per annum: not bad for the lowest grade of agricultural land, especially if you add the possibility of a bird hide which people will pay to visit. Yet in 1986, the United Kingdom was importing most of its thatching reed from Hungary, France, Romania, and Poland. The thought of Polish reed after the Chernobyl incident gives one visions of American tourists tucking into their cream teas in such places as Anne Hathaway's cottage beneath radioactive thatch! Even in France the reed industry seems scarcely better organized. The Grande Brière in southern Brittany, one of the largest reed-beds in Western Europe, imports thatching reed from the Camargue to supply roofs for its traditional buildings, which are perched on islands in the middle of the mighty reed swamp.

Customs and Excise in the United Kingdom does not separate reed from withies and bamboo in its calculations to assess imports, but the figure for all these materials imported in 1985 came to well over £2 million.[34] Home-grown reed costs less than imported supplies, is more appropriate as a local building material to reinforce a sense of place, and also creates winter employment in the countryside. There has been controversy concerning some English reed weakened by fertilizer leaching off adjacent fields. Hard oily reed, unaffected by nitrates, is slow-growing, and therefore slow to rot, once cut. By no means all English reed has been affected by nitrates, however, and the British Reed Growing Association is currently looking into the implementation of strict quality controls on home-grown reed. The solution, of course, is not to abandon the English reed industry, but to reduce our lavish use of fertilizer in the countryside. It is time that the reed industry, government, and conservationists worked together to create some major new reed-beds on marginal riverside land for which a conventional farming future seems uncertain. If so, we might allow the flood to lap over some river banks again, and the boom of the bittern may then be enjoyed in parts of the country where it has not been heard for hundreds of years.

## The Economic Value of Rushes

Another riverside plant, with potential for practical use, is the true bulrush, *Scirpus lacustris*, whose glossy jungle-green stems tower up out of rivers with large-scale exuberance more in character with

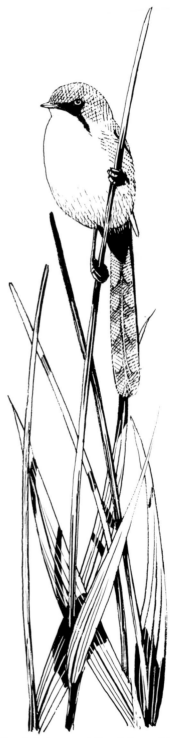

Bearded tit, a specialist of the reed beds.

Sweet flag, *Acorus calamus*, introduced by the Elizabethans, is still used in the perfumery business for its orange-peel fragrance. It is growing here in the moat at Harvington Hall, Worcestershire, where leaves were discovered in priest's holes.

tropical waters than with the demure flora of Northern Europe. Dredging has eliminated it from many streams, and I have often had to defend the retention of its clustering stands. Imagine my delight therefore when, in 1984, I received a letter from a Warwickshire chair-maker, pointing out that it was not only the wildlife which was imperilled by water authority activities, but also his livelihood. Neville Neale has been cutting rushes for his chair seats on the river Leam and the Warwickshire Avon since 1939, and he still has as many orders as he can meet, exporting some chairs to as far away as Australia. Now, he and his son cut the vigorous bulrush stands in high summer in places where they might otherwise be removed as an impediment to flood flows. This river weed, then, has a small but definite commercial value, and ratepayers' money is saved since the rush cutting (from which the plants ultimately recover) obviates the need to go in and dredge.

There are good reasons why this small example of co-operation between rush-harvesters and river managers might become more general. Visitors to Blickling

Neville Neale cutting rushes. River Avon, Warwickshire.

Hall in Norfolk and Hardwick Hall in Derbyshire walk upon woven-rush matting, which covers much of the floor of these great houses. At Bess of Hardwick's great Elizabethan palace of glass and stone, rushes have probably been strewn on the floors ever since the house was first built, as is indicated by some early coloured drawings. For all their Renaissance sophistication, sixteenth-century architects were still building with the materials of the local landscape; and in this spirit the National Trust has reopened the nearby stone quarry, unused since Bess's day, to reface Hardwick's magnificent main façade. Sadly, an unimaginative approach to river management has discouraged the use of another, much less expensive, local material. As long as memories serve, going back to the turn of the century, the Hardwick matting was made by old ladies in Bolsover out of locally gathered rushes. But when the last chatelaine of Hardwick, Evelyn, dowager duchess of Devonshire, opened the house to the public in 1947, because of the decline in the local industry, she had to order new matting from the Waveney rush industry in Suffolk, which still supplies the National Trust at Hardwick. Because of inadequate supplies of rush in the over-dredged rivers of East Anglia, however, this firm has been driven to import all its rushes from abroad. As a result, the cost of the product has been raised, and its repair is a major item of expense at Hardwick; meanwhile, up and down the rivers of England every summer, rush is cheerfully dredged out or mown by weed-cutting boats. Waveney Rush Industries, which provides considerable local employment, is expanding and is keen to find convenient supplies of rush to reduce its overheads. The Waveney rush-weave, which ranges from flooring to log baskets, carries with it a surely marketable appeal, combining the historic traditions of a great English house with the natural splendours of rushy English rivers.

## Water-meadows and Wetland Fodder Crops

There is an even more promising economic resource to be found along wild riversides, however. Because of the environmental benefits of winter flooding for birds, conservationists have long been interested in the practical restoration of water-meadows which, following their introduction in the sixteenth century, were the last word in sophisticated farming technology alongside many southern English rivers. Water was 'floated' along brim-full channels, as described by Thomas Hardy, 'on a plan so rectangular that on a fine day, they look like silver gridirons'.[35] This warmed up the land for the 'early bite', thereby allowing heavy stocking of sheep. Water-meadows reached their peak of popularity in the nineteenth century. There has been interest in their potential as fisheries, and trials in Wiltshire and New Zealand, reported in the *Farmers' Weekly* in 1979, explored the possibility of modern-style water-meadows as a way of increasing sheep stocks without the expense of nitrogen.[36] However, the cost of restoring and maintaining the channel systems has helped to ensure that water-meadows have not exactly swept back into fashion,[37] although, even here, the technology of simple silicon-chip sensors, switches, and pumps could take the labour out, and might just resurrect the traditional water-meadow as an up-to-date device of modern farming.

A more promising solution to the problem of achieving good grazing while retaining water over the land in winter may lie in a waterside weed which will perhaps become the wetland alternative fodder crop of the future. It already serves as such in the United States and East Germany. Reed canary grass, *Phalaris arundinacea,* which, to the layman, looks like a cross between a grass and a small reed, has long been grown in cottage gardens in its variegated form known as gardener's garters. Through long evolution, it has adapted well to wetlands, and studies at the Open University have shown that it responds well to fertilizer in wet conditions, yielding more fodder in such circumstances than rye-grass can produce on drained land.[38] Reed canary grass can be grown easily, and forms a compact turf for farm vehicles to drive over, although these may need to be modified if the land is very wet. It is palatable to cattle, and the only critical management required is regular cutting for forage, rather than continuous grazing. In this way, wet grassland for winter birds could be compatible with high fodder yields; and by tailoring the crop to the habitat, expensive land-drainage schemes to improve grassland could be rendered obsolete.

## Haymeadows

Far and away the commonest harvest of riverside land has always been grass for cattle, and it is hard to imagine that pasture will not remain so, particularly once surplus grain creeps back from many of the inappropriate riversides where we have planted it. What is no longer common and was never commonplace is the traditional management of grass for grazing and haymaking. North Meadow is a level field of 112 acres in north Wiltshire, which forms an island between the river Churn and the river Thames at Cricklade. It is still pastured in winter and mown for hay in summer by the local farmers, in a way that has been documented since the early Middle Ages, and was possibly begun when Alfred the Great founded Cricklade in 890.[39] North Meadow is Lammas land, which means that it is available for communal grazing from old Lammas Day, 12 August, in this case to any resident of Cricklade, who may pasture a maximum of ten head of horses or cattle until 12 February. After that the meadow reverts to private ownership for haymaking, the lots of hay being auctioned off at a pub in the town. The grazing is administered by the hayward—a confusing title since he has nothing to do with hay—on behalf of the court leet of Cricklade, whose officials, with titles including those of scavenger, constable, and ale-taster, sit down to dine every autumn with the due ceremony which they have observed since 1256. Every 12 August, Jim Marley, hayward and postman of Cricklade, opens the gate into North Meadow, and seventy-five head of horses and cattle thunder into the field with a lack of sense of ancient ritual which would do justice to a rodeo. Their tails whisk, and they stampede off into the wide open spaces of their new-found freedom. One year, thirty horses galloped away in the meadow, charging straight up to the end of its 100 acres and vanishing into the willow-shrouded distance. With obvious relish, the beasts then tuck into the green sward; but what they are eating is a great deal more than grass.

Wetlands such as Wicken are chess-board landscapes, and a chequer-board, for which the Latin word is *fritillus*, has even given its name to a butterfly and a flower of the wetlands. The marsh fritillary butterfly, a scarce beauty, splashed and chequered in marmalade and black, has been recorded at Cricklade. The great glory of North Meadow, however, is England's largest colony of snake's-head fritillary, a flower whose subtle freckling of purple and white epitomizes the qualities which Gerard Manley Hopkins admired in his poem 'Pied Beauty' when he wrote: 'All things counter, original, spare, strange.'[40] If the devil had a buttonhole, it would be of snake's-head fritillary. The flowers have a *fin-de-siècle* decadence, which Rennie Mackintosh captured in a 1915 water-colour, around whose title he extended the chequer-motif, and which was also sensed

Swifts and swallows swoop low over their insect harvest at haymaking.

by those who gave fritillaries their English country names: toad's head, frogcup, leper lily, and dead man's bells. Their slightly sinister beauty, so unexpected in an English meadow, is the essence of the fritillaries' glamour. It is appropriate that such a special flower should retain the last core of its distribution in England around the sources of the river Thames. At Easter, from Cricklade church, whose tower was built by John Dudley, the evil genius behind the coup which led to the brief rise and fall of Lady Jane Grey, the bells ring out across North Meadow. Above them rings the singing of up to ten breeding skylarks, registering, as Edward Thomas described it, 'in some keyless chamber of the brain'; while below, all around the stone markers placed across the meadow during or after 1824, to delineate the different hay plots, and laced with the gold of kingcups and dandelions, nod the innumerable purple bells. The wind, which is always keen across North Meadow, catches the flowers to reveal their crimson throats. As you walk out into the level spaces of the field, the 100-acre purple carpet intensifies. The flowers are never still, and the larks are never silent.

Three months later the meadow is mown for hay. The owners of the plots have 'Right of First Vesture', which means anything that falls to the sweep of the scythe. As the patient mower narrows the tawny grass creamed with meadowsweet and scorched dark with red-black burnet heads, the flowers of the meadow fall in long ribbons of colour behind the clattering blades. The papery hay-rattle seed pops and snaps among the fragrant mowings, and swifts and swallows swoop low in search of their insect harvest. Our modern machinery—the tractor, the binder, and the baler—may be less picturesque than scythes and shire horses, but they have not fundamentally altered that timeless sense of men working in the wide meadows which Constable painted and Edward Thomas wrote about, and which is the essence of July:

> Only the scent of woodbine and hay new-mown
> Travelled the road. In the field sloping down,
> Park-like, to where its willows showed the brook,
> Haymakers rested. The tosser lay forsook
> Out in the sun.[41]

What has changed in most places is the quality of the crop. Richard Mabey has described the sheer beauty of what has been surrendered for a greater yield of grass: 'and then, suddenly this brilliant field lapped with layers of colour and movement—yellow hay-rattle, red betony, purple knapweeds and orchids, the swaying cream umbels of pepper saxifrage, and butterflies so dense and vibrant above the flowers that it was hard to tell them from the heat-haze'.[42] No meadow is like any other. Even within a county, the subtle blend and mix varies with the character of the land. In Wales there are 'waxy-coloured mixtures of globe-flower and greater butterfly orchid'; whereas Teesdale meadows grow 'vivid purple mottlings of cranesbill and melancholy thistle'.[43] It is hard to realize that such splendours are not part of the cherished heritage of our most visited National Trust gardens. (I believe that the great tradition of English cottage gardening, refined to a sophisticated art by Gertrude Jekyll and Christopher Lloyd, actually harks back to the melting mixtures of permanent pasture.) Yet all this is is hay: stuff people roll in and feed to cows. Such was the universality of the sheer landscape quality which we have cheerfully abandoned. North Meadow at Cricklade now contains 80 per cent of all our English fritillaries. In the mid-1960s, the Nature Conservancy hesitated to designate the area a Site of Special Scientific Interest on account of its status as a common, and also because gypsies picked the flowers for market, and it was felt that there was nothing you could do about gypsies. Thus was the relative abundance and safety of other fritillary meadows taken for granted. By then, however, the new agricultural revolution was well under way, and everywhere old pasture began to be drained and ploughed for three-year rotational leys of uniformly green rye-grass. The reduction in seasonal flooding, which studies in Sweden have shown helps trigger the germination of fritillaries,[44] has also hastened their decline. Before 1930 fritillaries were present in 116 ten-kilometre squares in 27 counties, including Somerset, Devon, and Lincolnshire. By 1970 they had dwindled to 15 squares in 9 counties, and now they are limited to approximately 13 good sites in the country.[45] But that is just fritillaries. In 1983 the Nature Conservancy Council declared that only 3 per cent of flower-rich meadow turf had survived agricultural intensification since the War.[46]

What is the solution for this, one of the real prices we are paying for progress assumed to be for the general economic good? At Cricklade, purchase of the freehold by the Nature Conservancy Council has ensured protection, although one of the satisfactory things about North Meadow is the way it is still farmed on the council's behalf. A real economic value for hay has to be the best answer, if such can be found. A regime of grazing and haymaking is certainly the only way such flowery meadows will survive. One agricultural economist has actually suggested haymaking by town-dwellers starved of such rustic delights. One can just picture the smocks by Laura Ashley, the scythes from Habitat, and the sports cars parked down the lane! A more hard-headed solution for some meadows is already being tried at Cricklade: harvesting of seeds of wild flowers for gardens and for the re-establishment of native flowers elsewhere.[47]

Since 1982 up to 20 acres of North Meadow have been harvested by a seed-producer, who obtains clean samples of the wild-flower seed by sucking it out of the standing hay crop with the use of a tractor-drawn harvester, after which the grass is cut. Sales for this seed have boomed. During 1982–3 more than 3,000 kilograms of wild-flower seed was sold in the United Kingdom, with a retail value of around £100,000. This is all part of a horticultural boom, in which gardeners bought over 84 million packets of seed in 1985.[48] Wild-flower seed is certainly at the top end of the price range, selling well at between £35 and £50 per kilogram. The firm harvesting seed at Cricklade estimates a gross annual profit of a staggering £1,500 per acre, with the cost of the actual harvesting estimated at £160 per acre. This is five times the profit for wheat, for which a 'ploughing grant' was made available in 1952, specifically to encourage the removal of permanent pasture. Although there is bound to be limited potential for this exclusive commodity, no ceiling on the demand for wild-flower seed had been reached in 1986. For management reasons, the Nature Conservancy Council will allow only one-fifth of North Meadow to be harvested for seed, so that offers hope for other small meadows around the country. Gardeners and those creating public amenity areas would then be able to obtain their local county meadow mixes, which would further reinforce a sense of place. As more people buy wild-flower seed, it is certain that the price of the seed will come down, thereby further raising the ceiling of possible sales.

Over the past ten years, some very questionable European seed mixtures have been marketed as native English wild flowers; now that this has been virtually stopped, it would seem wrong to export our own seed. However, it may be that there is a foreign market among the anglophile gardeners of the eastern United States. The product would not then be wild-flower seed—they have their own appropriate American species—but the harvest of a haymeadow beside the Thames which has been managed since the days of King Alfred—the epitome of the Englishness of England, which we know has an emotional value both at home and abroad, and, consequently, if only we use our imagination, a hard commercial value as well.

There is also a powerful argument for maintaining unimproved pasture as a resource for the widest range of subtly different and still evolving varieties of grass, which may well be needed by future generations of plant-breeders. Some of our best Aberystwyth strains of modern grasses were developed by Sir George Stapledon from genotypes of native grasses which he found in old flood meadows. Stapledon also recommended broad-leaved herbs such as plantain, cat's-ear, and yarrow for their undoubted nutritional value, and burnet for its winter growth, which extends the grazing season.[49] 'The cattle came out of here, looking a picture,' says Jim Marley, the hayward of North Meadow. Many farmers will bear witness to the way in which beasts let loose into a freshly sown grass ley will make straight for the edges of the field, where the plough has been unable to reach the old weedy sward, so evidently tastier than the new grass, which must be the pasture equivalent of keg beer or sliced bread.

Agronomists working in the 1940s and 1950s, when many of the ground rules for modern grassland agriculture were being laid down, stressed the value of plants such as hay-rattle for their high mineral content.[50] German and Dutch researchers also emphasized the value of dandelions as a nutritional forage plant,[51] surpassed only by white clover, and containing high protein, amino acids, carotene, and major trace elements. The elimination of such plants has ensured a gain in bulk quantity of grass; but with that has come a greater reliance on fertilizers to sustain the modern varieties, together with a possible increase in vets' bills.

The landscape quality of the countryside is also diminished by the disappearance of such everyday flowers as dandelions from our brand new leys of unmitigated green. Their absence, unnoticed until, with a shock of surprise, you come upon a dandelion field and realize that this sunny, unexceptional sight has indeed become the exception, is as sad in its way as the dwindling of fritillaries. Around Stratford, where the country name for dandelion clocks was 'chimney-sweepers', their ordinary gold is reduced largely to grimy road verges, a last reminder of the Maytime exuberance which prompted Shakespeare's memorable elegy:

> Fear no more the heat of the sun,
> Nor the furious winter's rages;
> Thou thy worldly task hast done,
> Home art gone, and ta'en they wages;
> Golden lads and girls all must
> As chimney-sweepers, come to dust.[52]

The fifty beef cattle belonging to Harold Ody of Clattinger Farm, between Cricklade and the source of the Thames, graze off an hors-d'oeuvre which includes green-winged orchids, dandelions, pale cowslips, and even the dusky bells of fritillaries. For Clattinger (whose name, pronounced Clattinjer, means 'eel trap') is in the parish of Oaksey, where fritillaries were once so abundant that the local term for them was 'Oaksey lilies'. Harold Ody's farm is everything which a riverside landscape ought to be. Massive hedges of embowering willows, which were planted by the tall, nineteenth-century incumbent of Clattinger who rejoiced in the name of 'Withy Pole Hawkins', shelter both cattle and farmhouse from the winter winds. Sloping down to the clear waters of the Swill Brook are 154 acres of haymeadow, a whole farm of wild flowers, which flourish on land as well kept as Mrs Ody's impeccable kitchen garden. 'I've ordered it,' says Harold Ody of the land which has been in his family since 1918.

Content to make a decent living, Harold Ody has not over-extended himself with loans from banks in order to maximize all possible inputs to his farm. He has never put out money on fertilizers, but simply recycles farm manure on to his fields; and his refusal to lay out on the expensive equipment required for making silage is one reason why the meadows at Clattinger Farm survive as a nationally important treasure. The general decline in permanent pasture has

responded to a simple financial equation: high grain prices, upheld by the Common Agricultural Policy + publicly funded drainage schemes + 50 per cent underdrainage grants = higher profits + higher land values. This artificial economic base underpins a wasteful, inefficient approach to husbandry. The real cost of ploughing, draining, and fertilizing every inch of land from riverside to hilltop to create landscapes like that at Orton-on-the-Hill is unlikely to have been incurred without generous state aid. Of course the land can be made to yield its maximum physical tonnage, but at what hidden cost?

Sir Richard Body, farmer and Tory MP for a farming constituency, made the point thus:

To grow bananas on the side of Ben Nevis would be totally absurd, but technically it could be done. . . . Huge glasshouses would be necessary and constructed with elaborate support to withstand storms and snow, and then heated to an inordinate degree. Special varieties of bananas would have to be evolved. . . . When the bananas came to be sold, the economic price might be £3.00 or £4.00 each, and then the special pleading would begin. 'Buy British', we would be told; 'every pound spent on a home-grown banana is a pound saved on the balance of payments'.[55]

This parable of absurdity makes a valid point about the millions of pounds run up as a result of a European agricultural policy which takes little account of the differing climates of its member States, and has thus encouraged the production of massive grain surpluses on our wet, westerly Atlantic island. Tragically, farmers such as Mr Powell, who have been lured by government policy into spending every penny they have upon agricultural intensification, will now have difficulty in finding the capital to de-escalate and diversify their farming practices.

In 1985 and 1986, with major land-drainage grants cut and the price of wheat promising to fall, some farmers were still carrying out land drainage to promote cereal production, which, with the enforcement of dairy quotas, continued, at least for the time being, to offer a good return. In 1985 a Staffordshire farmer engaged a local drainage contractor to straighten out the river Blithe where it flowed through his land, and also to strip out every hedge and all the trees, which included mighty oaks and willows, from the adjacent fields. Such are our planning laws that a householder overlooking the Blithe valley requires planning permission, which may be withheld, to put in a dormer window or erect a fence higher than 2 metres around the edge of his garden, while it is possible for a farmer to radically transform the total landscape for the worse in the name of agriculture. Furthermore, because land-drainage by-laws naturally enough prohibit only the worsening of flood conditions, the water authority, which had spent public money taking meticulous care of adjacent reaches of the same river, was able to do little more than watch the farmer fell almost every tree along the river as he turned his section of one of the loveliest valleys in the Midlands into a single field of wheat.

*Recreation and Tourism*

On neighbouring land, where ancient hedges still sweep down to a wandering alder-shaded stream, the Blithe was systematically engineered in 1986 to reduce flooding without any sacrifice of ecological value. Here, the well-managed river mirrors a managed, yet beautiful, valley farm. John Miatt, who works this attractive and productive landscape, is no laggard. Indeed, this leading figure in the local branch of the National Farmers' Union, makes a wise point. He receives a good supplementary income by letting out his converted farm buildings to fishermen and holiday-makers. The pretty valley of the Blithe is what they come for. Similarly, Richard Porter, in Herefordshire, has, for a handsome sum, sold an old barn to a retired businessman for conversion into a second home. In the South-East, where every oast-house conceals a man of letters or an advertising executive, a Sussex barn was reported in 1986 as having been converted into a dozen homes, selling at £135,000 each.[56] There are unlikely to be such enthusiastic takers for second homes overlooking Orton-on-the-Hill.

Fishing, our largest national participation sport, earns many hundreds of thousands of pounds each year. In 1985 a prime salmon beat cost £500,000;[57] and the many miles of humbler, coarse-fishing rivers yield steady rents from a growing army of approximately four million anglers. On some farms, shooting can also be a prime earner, and the British Association for Shooting and Conservation gives advice to landowners on pond and wetland management, as well as the creation of game cover along river corridors. Snipe, a classic favourite among sportsmen, has been a typical casualty of agricultural drainage.

The way in which intensive agriculture can eliminate a recreational resource for sportsmen and naturalists alike was well illustrated in the spring of 1985, when the Wildlife and Countryside (Amendment) Bill had a stormy passage through Parliament. A critical clause proposing that the Ministry of Agriculture has a 'duty to further wildlife conservation' was removed from the bill at the eleventh hour. One of the strongest opponents of this clause, both at the second reading and in committee, was Sir John Farr, MP for the Leicestershire constituency of Market Harborough and a keen fox-hunter.[58] It soon became clear that the sportsman's vote against an integrated directive for agriculture might also prove to be an own goal against the interests of the chase. Leicestershire, pre-eminent as a fox-hunting county, had lost 46 per cent of its permanent grassland since the War, largely to arable farming, with an accompanying massive loss of hedgerows, trees, and copses. Not surprisingly, with landscapes like that at Orton dictating the new shape of the shires, foxes were being squeezed out, along with all other wildlife. That spring, the Leicestershire Hunts reduced their hunting days, in part because of the shortage of foxes, while Lincolnshire huntsmen were arriving in Leicestershire in pursuit of animals which had fled the eastern prairies of their own county. The fox is an adaptable beast, however, and as pressure on the countryside has become ever more relentless, foxes have become a much observed phenomenon of towns and cities. In 1985 the Leicester City Wildlife Project published details of a survey which showed that over 250 foxes were living in Leicester itself.[59] Could it be that the huntsmen of Leicestershire will be reduced to galloping through traffic-lights in pursuit of their quarry? In Lincolnshire, one spectacular kill was made in 1981 on the platform of Gainsborough railway station, to the astonishment and consternation of passengers alighting from the London Express.

Wetlands may offer a major attraction in themselves, as part of our expanding tourist industry, and with careful management, their wildlife could be enjoyed without undue disturbance by those willing to pay for the pleasure of admiring it. Such places are ambassadors for the rest of the natural system. North Meadow at Cricklade receives between 2,000 and 3,000 visitors in the flowering season. Wicken Fen, one of the National Trust's earliest acquisitions, may only now be coming into its own, as an asset equal in value to the trust's stately homes. The National Trust, which was founded originally by Octavia Hill to preserve commons and wild places, is fast rediscovering its roots, with a fresh emphasis on its countryside land-holdings, echoed by increasing interest in these places from the public. Martin Mere, near Southport, a wetland restored after years as poor summer grazing and a farmyard used as a rubbish dump in the 1970s, is now managed for its birdlife. In winter, the sky above the ploughed Lancashire Levels, won from the swamp after centuries of patient drainage, is now clamorous with thousands of pink-footed geese as, in V-formation, they home in towards their reserve. In the wake of the wildfowl have come coach-loads of visitors. So successful has the 'cash crop' of spectacular wetland wildlife been, attracting an annual average of 185,000 people through the turnstiles, that the Wildfowl Trust has been able to extend the habitat for the birds, thereby turning back the clock, as they inundate the land beneath the waters of the western plain.

In 1985 the tourist industry in the United Kingdom had an annual turnover of around £14 billion, which is more than the whole of our agricultural sector; and in 1984 it created approximately 70,000 new jobs. One of the things which tourists come to see is the English countryside, enshrined in our culture and famous throughout the world. Its special character, compared, say, to that of Australia, Canada, or Africa, which have far finer wildernesses than England, is a quality of intricacy and variety, resulting from the fact that ours is a landscape of long occupation. This sense of place, subtly distinguishing each English parish and county from the next, has long been marketed by the tourist industry as one of our special assets. We have Shakespeare country, Constable country, Wordsworth country, and, with the advent of television, Herriot country. While one branch of the Government promotes tourism, however, other branches, such as the Ministry of Agriculture, promote the steady erosion of the landscape which is one of the basic resources of the tourist industry. In the late 1970s I well remember finding plenty of money available in the water industry for creating picnic sites beside rivers, while no money or staff were available to

modify the concreting and canalizing which another branch of the same organization was carrying out on the same rivers. This is symptomatic for the whole
countryside, for which an effective land-use planning policy does not exist.

Beside the river Severn at Ironbridge is an open-air museum, which typifies
the most up-to-date attractions promoted by the British Tourist Board. Alongside
one brook, running down to the Severn, is a reconstructed village, in which the
main exhibits are actually the staff, dressed in period costume. Such places are
imaginatively designed, and do a roaring trade. There is nothing wrong with
them in themselves. What concerns me is the setting for these magnificent
tableaux, with their top-hatted gentlemen offering pony rides and their crinolined
ladies tending cottage gardens and demonstrating country crafts. Should the
unwary tourist creep behind these stage sets evoking a landscape in which man
and nature are in harmony, he will be brought up against the fast-vanishing
reality. The Ironbridge museum is in Shropshire, a county enshrined in the poetry
of Housman as the epitome of unspoiled England:

> In valleys of springs of rivers,
> By Onny and Teme and Clun,
> The Country for easy livers,
> The quietest under the sun.[60]

In 1985 the *Ecological Flora of the Shropshire Region* was published, establishing a clear picture of what had been happening to those 'springs of rivers'
since Housman's time.[61] Modern surveys, combined with a careful study of old
records, allowed a comparison to be made between Shropshire's flora before
1913 and after 1969. In considering plants of acid bogs and wet heaths, it was
found that one species had already become extinct, while on most sites the range
of species had been halved. In nutrient-poor lakes and pools in Shropshire, where
up to twelve different species had previously been found, the score for the best
site was now four, while no other site had more than two species. Such losses
are the direct result of drainage and the leaching of nutrients from farmland.
The absence of such flowers does not make or break a landscape, but it does act
as a good litmus paper to indicate a general picture of decline. A rather less
subtle indication was the recent edition of the *Guinness Book of Records*, in
which Shropshire boasted England's largest field. The naturalist must now travel
through impoverished landscapes in order to see quite commonplace examples
of his local flora and fauna, which often survive only in nature reserves or at
Sites of Special Scientific Interest: the equivalent of a holiday-maker's journey to
a recreational centre, where he can dream about the past glories of the English
landscape. It is an extraordinary fact that the more the *reality* of our complex
historic landscape disappears, the more we take refuge in elaborate rural and
historical fantasies in our leisure time. These can range from the animated
waxworks of Queen Victoria at Windsor, promoted by the British Tourist Board

as the ultimate attraction in 1984, to yet another television series skilfully filmed to reveal a happy and healthy wildlife, disporting itself among landscapes of idyllic beauty.

Images of traditional countryside are indeed the standard vocabulary for advertising and marketing the products of drainage and high-gear agriculture which directly destroy that countryside. Take a look along the shelves of the local supermarket, or simply open the door of the fridge. On cheese, butter, and bread wrappers, the hills are never treeless, cottages are invariably thatched, and rivers always wind. The token contented cow grazes upon flowery meadows, whereas in reality she is often fed on grain, not to mention her own butter. My favourite is a yoghurt pot, on to whose painted landscape of green pastures and water-meadows the artist has set beside the brim-full stream a thatched water-mill.

## The Misuse of Peat

Among all the harvests of drainage, however, one stands out as epitomizing our lack of comprehension about what we are actually buying. It is peat. The black gold of the wetlands, it is used in commercial horticulture, and above all by gardeners for grow-bags and for spreading on their flower-beds. Peat is the accumulated and compressed remains of pickled plant material, a kind of wetland compost, which, because of the waterlogged conditions in which it is found, does not decay, but over sufficient time, actually grows, as layer upon layer of dying leaf and stem build upon its surface. About 90 million tonnes of the world's peat are excavated every year, far beyond the rate at which fresh deposits are formed. For this reason, peat should be regarded as a non-renewable resource.[62] In the United Kingdom, it is estimated that we use around 2.5 million cubic metres of peat each year, some of which is imported from Russia and Ireland to be mixed with our own peat dug from reserves in Yorkshire, Somerset, Cumbria, Lancashire, and East Anglia. In commercial terms there are two kinds of peat, both extracted from wetlands of great ecological interest: sedge peat, formed in freshwater swamps, and sphagnum peat, which comprises the remains of sphagnum moss, and for which major sources of supply are natural systems of outstanding fascination and beauty known as lowland raised mires. Such were the famous Lancashire mosses, largely destroyed now through drainage and peat cutting. Still with extensive peat deposits, but vanishing before our eyes, are the magnificent mires of south Yorkshire: Thorne Waste and Hatfield Chase.[63]

Hatfield Chase, or Hatfield Moors, lies, like the inner heart of an onion, ringed by a concentric system of dykes, within the drained and ploughed fen known as the level of Hatfield Chase. With its low horizon of trees, always beyond the fields of wheat, the Chase taunts the traveller, as all roads seem to circle round it, keeping a respectful distance. It takes a real explorer to find his way over the single rusting bridge, past wartime signs announcing that trespassers will be 'arrested and prosecuted', and so, finally, on to one of England's finest and

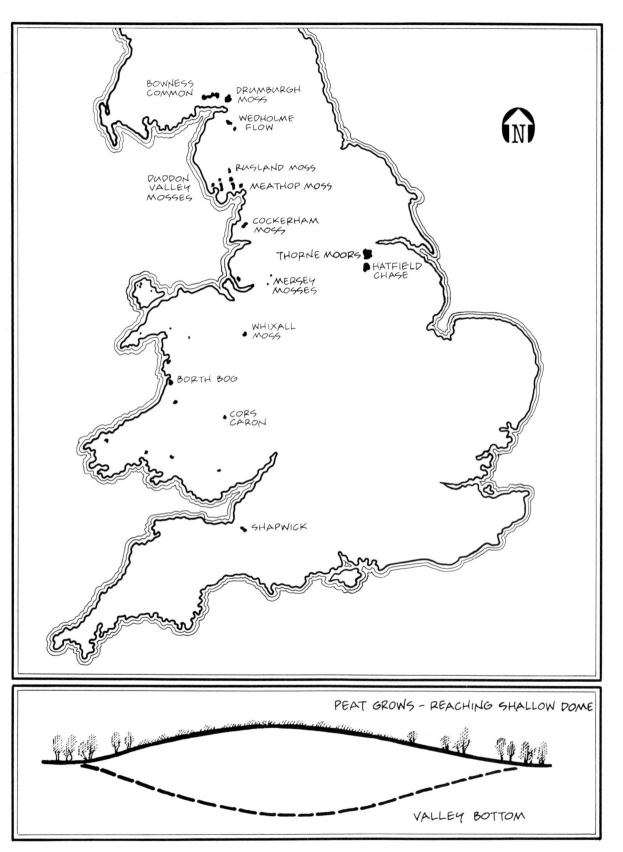

Lowland raised mires in England and Wales.

largest wildernesses: 3,500 acres of moss and birch and pine. This was the lowest and deepest centre of the fen, and for this reason the moss litter has patiently grown over the past 3,000 years, to form the immense shallow-domed pudding of peat which comprises a classic raised mire, whose paradoxical topography, rising to a crown at its very deepest centre, has defeated all serious attempts at drainage from Vermuyden to this day. Hatfield Chase resembles the wildest and most exciting heathy common, in whose centre survives a farmhouse on the site of the hermitage built by Billy Lindom, a medieval holy man and reputed miracle-worker. Among the glittering ranks of silver birch, from which project the creamy brackets of polypore and tinder fungus, the mysterious nightjar churrs to his mate. Forty-five breeding pairs were recorded here in 1986. Bog rosemary, *Andromeda polifolia,* much coveted by alpine gardeners, actually grows wild on Hatfield Chase, in some places forming pink and silver hummocks up to 200 yards wide. In spring the peat is white with a blizzard of cotton-grass, and the cuckoo sings all day. In late summer, the black and gold darter dragonflies, *Sympetrum danae,* sun themselves on the heather, and the scarce large heath butterflies feast on nectar from the coconut-scented gorse.

But the unforgettable spectacle which Hatfield Chase presents in the late 1980s, a sight which makes visitors' jaws sag visibly, is not so much its natural splendour as the ruin which man is inflicting upon it. Behind a discreet screen of wind-scorched Leyland's cypress are the shabby huts, tramlines, and stacked bales of the peat works; and beyond them extends 2,000 acres of totally stripped-out, black, gleaming peat: an area four times as large as Wicken Fen, equivalent to that of a sizeable town. The remaining 1,000 acres of vegetation, where surviving trees, birds, and butterflies still hold on, is also scheduled for destruction. The shifting light picks over the great sweep of the clean peat, devoid of even a single bobtail of cotton-grass, as the machines, looming out on the moor like space-age dinosaurs, pick over the raised mire, systematically dismembering it until there is nothing left. Hatfield Chase took 3,000 years to evolve, and if properly managed, would continue to develop as a living system. Now, in perhaps ten years or less, it will be gone, taken away bag by bag, crumb by crumb, for brief summer crops of tomatoes in our greenhouses, and to be scattered on people's rockeries, and even on their garden conservation areas. For years, I actually used peat in planting schemes for the enhancement of wetland habitat. The ruining of the raised mires is a symbol not only of greed, but, more profoundly, of our ignorance of the consequences of our own actions. Gardening is a supremely life-enhancing occupation, in its way a celebration of our links with the natural world. We laugh now at eighteenth-century botanists such as Roscoe and Banks who founded botanic gardens while they promoted the draining of Chat Moss and the Lincolnshire fens. But our own destruction of places such as Hatfield Chase is the ecological equivalent of knocking down a cathedral and using the dust to line the garden path.

As so often happens, new technology has been the key to maximum

The glittering ranks of birch which dominate a classic raised mire. Hatfield Chase, South Yorkshire.

exploitation of natural systems. The first stage was to drain the mires sufficiently to allow them to be worked. Then came peat-digging machines, which enhanced the peat industry's appetite for land. The situation is comparable to that found in whaling, in which capital is tied up in ships and machines, which must then be kept busy in order to justify the original expenditure on them. Until the late 1970s, even the machines used at Hatfield Chase harvested the peat in a relatively agreeable manner, cutting 10-yard-wide strips across the mire and allowing adjacent vegetation to recover. Now, a thin skim is milled clean across the whole system, leaving much less of the vulnerable and isolated ecosystem to survive. As with high-gear farming, employment has been reduced through the mechanical intensification of peat winning. It would be naïve to sentimentalize the old backbreaking misery of manual peat cutting, but it is just possible, given the depth of raised mires (now reduced from around 20 to approximately 10 feet deep in the case of Hatfield), that low-level mechanical harvesting of peat could be carried out in a way which does not entirely eliminate the ecosystem of the mire. The potential of the peat resource in terms of biochemical products from peat extracts may also be far more important than its use as a dressing for gardens.

Certainly some management, and even harvesting, of peat creates the most diverse habitat. In 1982 the fen violet, *Viola persificolia*, reappeared at Wicken for the first time since 1916, as a direct result of disturbance to the ground.[64] This plant, which was also recorded at Hatfield in the 1960s, depends upon the conditions created by low-technology peat cutting for its survival: and an imaginative plan is afoot at Wicken to encourage the visiting public to dig their own peat by hand, thereby personally promoting the survival of one of our rarest wild flowers. With proper management, Hatfield Chase and the neighbouring and even larger Thorne Waste (currently threatened with destruction for peat extraction, except for 200 acres bought at great public expense by the Nature Conservancy Council) could also be the jewels in the crown of what is a potentially exciting tourist area, already used for cycling holidays and badly in need of an injection of capital, which the tourist industry would bring. Risley Moss, in Lancashire, a tiny scrap of surviving raised mire, by contrast with these south Yorkshire systems, has been very successfully developed as an educational attraction.

Planning permission for peat extraction on most of our raised mires was granted in the 1940s and 1950s, and it is now difficult to reverse. The Nature Conservancy Council has been faced with the prospect of horse-dealing with the peat companies, with offers to trade off some peat lands against others. So precious are these places that this is rather like saying that you can knock down Canterbury Cathedral if you allow us to buy Westminster Abbey. A far more satisfactory solution would be for the gardening public, not to mention publicly accountable landscape offices, to cease using the stuff. Commercial horticulture is starting to turn away from peat on practical grounds, in favour of artificial soils such as rock fibre and hydroponics; and there is no doubt that the garden-

An environmentally catastrophic consequence of the gardening trade: 2,500 acres of unique habitat levelled flat to provide horticultural peat. Hatfield Chase.

centre boom remains a major factor in the multi-million-pound peat industry. There are satisfactory alternatives for gardeners. Manure and compost were generally used before peat became easily available. Chopped bark is now on sale, and British Earthworm Technology at Cambridge has developed a light odourless humus as a by-product of chicken manure.

## Gravel Pits and Mining Subsidence

Peat working is really a specialized kind of open-cast mining, and the fact that the Norfolk Broads are actually the flooded remains of medieval peat cutting points to the exciting opportunities which all kinds of mineral extraction open up for the creation of new wetlands. This does not mean to say that a lake resulting from worked-out peat digging would be adequate compensation for the loss of so scarce a natural system as the raised mire of Hatfield Chase.

Coal mining, too, offers our generation an outstanding opportunity to create major new wetlands to bequeath to our descendants. In 1985, the National Coal Board envisaged a deep-mine output of 90 million tonnes of coal by 1990; and in the same year the Council for the Protection of Rural England calculated that open-cast mining was likely to take place over an area roughly equivalent to the size of Bedfordshire by the end of the century. The pottery-making town of Armitage, home of the British lavatory bowl, straddles a low hill beside the river Trent in Staffordshire. In 1986, £1 million of public money was spent by the National Coal Board on embanking and pumping 150 acres of farmland which would otherwise have sunk beneath the river level due to deep mining taking place beneath. The option of buying out the landowners and abandoning the land to the river was not adopted, despite the fact that it would have cost less than the protection option. As the massive engineering machinery moved in to prop up the fields in the interests of agricultural production which was every-where creating surpluses, the Coal Board and the water authority were unable to agree on who should maintain the banks and the pumps in perpetuity. The National Coal Board rightly stated that it was not its duty to look after drainage installations, and the water authority, equally correctly, pointed out that it was no longer expected to promote agriculture with the same vigour as it had hitherto. It is now clear that the option of abandoning agricultural land subject to mining subsidence should automatically be given far greater consideration than it has been over the past 40 years. Abandonment of land in such circumstances during the days of agricultural depression earlier in the century has bequeathed us such fine nature reserves as Leighton Moss in Lancashire, Stodmarsh in Kent, and the grim grand wetlands of the Ince Flashes at Wigan. The National Coal Board generally prefers to buy the land affected by its operations, and this would tie in with the abandonment option, which it is already adopting in the North-East. One problem is that structure and mineral plans of many county councils assume that restoration to agriculture is of paramount importance, while the agents for the Coal Board are most commonly either the Ministry of Agriculture or the Forestry Commission, neither of which is in all cases enthusiastically predisposed to projects of wetland creation. Another difficulty is the lack of any overall restoration policy for a single valley, let alone a strategic national plan.

The high politics of holes has been played out in a manner which has everything to do with the expedient solution of immediate problems, rather than long-term environmental planning. As with other so-called 'wastes', one man's hole for tipping is another man's habitat. The spectacle of marsh orchids being buried by old bed-springs and broken tarmac on a Staffordshire wetland in 1985 was a consequence of the district council concerned failing to recognize what was a fine relict fen habitat when they granted the initial permission for filling. While some holes, especially near cities, are in great demand for fill, there has been a reduction of available non-polluting fill material, especially pulverized fuel ash, due to the scaling-down of coal-fired power-stations. In the case of open-cast mining, it is generally difficult to calculate the exact balance of fill required until late in restoration, and this can lead to the costly importing of spoil. An early decision to allow valley bottoms of such areas to go under water can sometimes save that expense and result in a fine habitat into the bargain.[65]

Marsh orchids, gradually being buried under bedsprings. The opportunities which wetlands present as holes for dumping rubbish can lead to their destruction. Four Ashes, Staffordshire.

## *A Thoughtful Economic Policy for Drainage Needed*

It is clear that defence of attractive riversides and wetlands need not depend only upon their contribution to our quality of life. A drainage policy based on previously held assumptions concerning its beneficial contribution to progress, regardless of all other costs, can now be seen to be both wasteful and inefficient. This is not to say that all river management could, or should, be stopped immediately, or that all our valleys should be given back to the rivers. What is likely is that some riverside land previously drained to yield conventional crops will continue to contribute to the 6.4 million acres (2.6 million hectares) of farmland which, in 1986, William Waldegrave, Minister for Environment, estimated would be surplus to agricultural requirements by the year 2000, due to over-production of food and because modern farming yields a far greater harvest per acre than it did in the past.[66] It would be naïve to expect such wetland crops as cricket-bat willows, rushes, and wild flowers to provide an instant and unlimited economic solution for every riverside landowner. But if all the 'alternative' uses of wetlands, especially tourism and timber production, are combined within the framework of an approach to agriculture which emphasizes low inputs and diversity, they do provide pointers to a more innovative and thoughtful approach to wetlands than previous government policy, which encouraged farmers simply to contact their water authority and request a drainage scheme. In 1989 the Government plans to start the first stage of its Set Aside proposals, whereby farmers are paid between £60 and £80 an acre to take part of their land out of food production, either to lie fallow or to support alternative crops or timber. Under such a scheme the opportunities to push intensive agriculture back from many of our riversides, and even promote new wetlands along the river corridors, will need to be imaginatively considered.

While some innovators who explore alternative uses for undrained land will not survive, others will lead the way for a new generation of farmers, who may find an economic value in the undrained corners of their land. None the less, while farmers face what could be the worst agricultural recession this century, there are bound to be some landowners who will go bankrupt. While this is socially undesirable, it is well to be rid of the fallacy that such personal tragedies are automatically a disaster for the landscape. Records show that the farming depression of the 1920s and 1930s was an outstanding period for breeding wetland wildlife.

In 1984–5 the Government, recognizing the need for a change in emphasis, directed that urban flood alleviation take precedence over agricultural drainage. But, as we shall see in the last chapter, with local political control of land drainage still remaining firmly in the hands of influential farmers, environmentalists should not be too optimistic about effective and radical changes taking place everywhere. The need and the will to tame the flood lie deep in the heart of agriculture, and farmers are well aware that they have in their hands a neat knot of issues concerning the fate of the man in the flooded street, as well as the profit of the man in the undrained field. This is a knot which their opponents will have difficulty in untying. Rivers do not conveniently cease to flow and flood at the point where urban interests give way to agricultural land. It is always entertaining to hear agricultural drainage men dreaming up epithets such as 'flood alleviation' to give a more acceptable image to their activities. 'Please do all you can', announced the Association of Drainage Authorities, following the major axing of agricultural drainage by the Government in 1985, 'to spread the word that land drainage means safety from flooding and starvation'(!)[67]

Furthermore, agriculture has been here before. It has weathered depressions and survived, because the food which it produces is a basic requirement for our survival. Once the confusions and excesses of the Common Agricultural Policy have been remedied, it is possible that a call for more food production will begin all over again. This political and economic process is what has made—and can break—the landscape, which is also a precious national resource. Farmers have always been adaptable, and they will respond to the signals which they receive from Government. Just as river engineering now includes an environmental brief, let us hope that future economic policies will be thought out in such a way that they dictate the evolution, rather than the destruction, of our countryside. As we have seen, some of our richest landscapes are the product of resourceful management with economic aims in mind.

The false idea that every damp parcel of land can be squeezed dry and that every river can be made to stay straight is a good analogy to the equally fallacious, but still cherished, belief that on our small planet of limited resources, unlimited economic growth is either possible or desirable. Ironically, it has not been upon the conservationists, but upon the farmers, that the lending banks have been foreclosing in the late 1980s. Those who thought that the flood could be tamed entirely have been brought up against the limits to growth which we must all accept if life is going to be worth living in the future.

# 6

# *Civilizing the Rivers*
## *The New Approach to River Management*

I do not know much about gods; but I think that the river
Is a strong brown god—sullen, untamed and intractable,
Patient to some degree, at first recognised as a frontier;
Useful, untrustworthy, as a conveyor of commerce;
Then only a problem confronting the builder of bridges.
The problem once solved, the brown god is almost forgotten
By the dwellers in cities—ever, however, implacable,
Keeping his seasons and rages, destroyer, reminder
Of what men choose to forget. Unhonoured, unpropitiated
By worshippers of the machine, but waiting, watching and waiting.

THUS wrote T. S. Eliot in the *Four Quartets*.[1] That the river is always watching and waiting for its moment to flood is something which conservationists occasionally forget, but which river engineers can never afford to ignore. What the drainage men and the 'worshippers of the machine' have sometimes chosen to forget, however, is the paradox that, while rivers are potentially death-dealing, they are also supremely life-enhancing. Since the water authorities were set up in 1974 with a duty to manage rivers, an attempt to balance these two conflicting characteristics has begun to occupy those who share rivers as a boundary, as a potential flood problem, and as a unique environmental resource.

In the past, rivers were recognized by mankind simply as frontiers, symbols of division between warring tribes. Lest we assume that such rivers of blood are just a thing of the past, it is worth remembering a few contemporary examples. The Shaht al Araf waterway, an engineered river at the confluence of the Tigris and the Euphrates, where civilization began, is now a strategic divide in one of the most barbaric wars ever fought. The river Blackwater, which for part of its length forms the watery gulf between northern and southern Ireland, offers a physical dimension to that expression of concord 'to build a bridge', since many of its bridges have actually been blown up as a result of the present Irish troubles.

Happily, in our own times, the rivers of England and Wales, the epitome not of division but of tranquillity, have also become a very special test case for

Some of the agreeable and useful things which wetlands can provide. The chair is seated with rushes. The basket, hamper, cricket-bat, and aspirins are all derived from willow. Leeches in the leech jar are still used in hospitals, and the wild flowers are proving a lucrative crop for the seed merchants.

modern co-operative countryside management. The sheer undeniable beauty of river landscapes has been their own best advocate in bringing about a common realization that a new kind of bridge building in river management is needed, involving engineers, digger-drivers, farmers, and conservationists, to ensure that, while the risks of flooding are reduced, the wildlife and the wild quality of the rivers are preserved and even enhanced. With this realization came another discovery: that, subject to some critical qualifications concerning cases in which any drainage would be environmentally inappropriate, the whole, hitherto destructive, process of river engineering could now be turned on its head. By working to a brief which takes full account of the life-enhancing potential of a river, rather than its narrow definition as a drain, the erstwhile destroyer can now become a creator. It was recognized that machines which previously straightened every bend, filled every pond, and felled every tree could, at very little extra expense and at no sacrifice to drainage efficiency, actually extend riverine habitats for dragonflies, kingfishers, and otters, not to mention add to the sum of human happiness for children with jam jars, walkers on riverside footpaths, and landowners out for an evening canter or in pursuit of duck and pheasant.

## The Reasons for River Abuse

To explain why the reverse has happened and why river after river has been scoured out and reduced to a treeless drain running straight to the horizon, it is first necessary to understand how landscape features and wildlife habitats of a river, if allowed to develop totally unchecked, are often the cause of flooding. The best way to understand the potential conflict is to look not at a photograph, but at Millais's painting of the drowning Ophelia (which also demonstrates the special place which rivers hold not only in our ecological system, but in our culture). Engineers always enjoy this particular illustration of their dilemma, since they like to imagine that the drowning maiden, complete with her posy of botanical specimens, is the lady from the local naturalists' trust, flung into the river after a particularly exasperating altercation over the conflicting aims of river management! To observe the riverine setting for his painting, Millais set up his easel near Ewell in Surrey, beside the Hog's Mill river, where he was much beleaguered by swans and midges. He accurately depicts flood-wrack on the bank at the right-hand corner of the painting. This is scarcely surprising, since all the habitats which the artist portrays so lovingly would be certain to give modern river engineers some headaches. The pollard willow, complete with germinating seedlings in its crown and a robin perched on it, leans perilously out into the stream. Left there for long, it will undoubtedly fall into the river and cause a blockage. The dogrose, splashed over with white July flowers and creating welcome cover under which birds and animals can scuttle for safety, will be sure to catch all the dead dogs and other debris to come bobbing down in the next winter flood, and make that blockage even worse. To the right of it

*Top.* Millais's painting of Ophelia. The leaning willow, dogrose, eroding bank, and water-weeds are equally a delight to the naturalist and a headache for the engineer. Photo: Tate Gallery. *Bottom.* The machine which was previously the destroyer can now be the creator of riverside habitat. Here water forget-me-not is being transplanted as part of a sympathetic engineering scheme.

is a vertical river cliff, whose unstable eroding surface has provided a fresh seed-bed for the biennial teasel, but which is likely to continue to slip into the stream, despite the binding roots of the purple loosestrife on its top. Beside Ophelia's head, the ever invasive forget-me-not and tall clumps of branched bur reed, essential for the survival of many insects and fish, grow out into the stream, trapping silt and clogging the watercourse. Finally, to ensure a flash-flood with the advent of August thunderstorms, the mossy webs of crowfoot floating in the foreground could raise the mean summer water-level of the stream by as much as a metre. You can be sure that the Thames Water Authority would never allow the Hog's Mill river to get into such a dangerously picturesque state today.

Taking the removal of such obstacles, together with islands, sand-bars, and meanders which slow down the flow, to its logical conclusion, managers of rivers began to develop the nearest they could achieve to a U-shaped section for channels, since this economically maximizes water flow without creating erosion or deposition. The main point of the exercise was to get water off the land and down to the sea as fast as possible at periods of high rainfall, by speeding it through the watercourse at a maximum number of cubic metres (cumex) of flood-water per second. In manipulating watercourses in this way, engineers have based their understanding of river behaviour on regime theory, which was first formulated and refined between the 1890s and the 1930s in the Punjab, by engineers working in the service of the British Empire on the construction and monitoring of monsoon drains, in climatic and geological conditions entirely different from those which affect rivers here.[2] The decision to carve out straight U-shaped channels is generally based on the application of an equation known as Manning's formula.

One of the fundamental variables in Manning's formula which must be measurable, therefore, if the formula is to be strictly applied, is the roughness of the channel through which the water flows. To measure precisely the roughness of the bed and sides of so living and volatile a thing as a river is like trying to catch a rainbow. A river's inside measurements depend upon the vegetation growing in it. This in turn depends upon the distribution of plant clumps, their density, height, shape, and flexibility, all of which vary over a short reach, and virtually from day to day in the growing season. Furthermore, while the vertical growth of plants such as bulrush and bur reed increase the roughness factor, it is known that the streamlining of flexible leaves and stems of some water plants actually smooths the channel, and certainly does nothing to increase friction at high flows. Very little work has been done in this country on the effects of vegetation on channel capacity, and the roughness caused by trees has been the least studied of all. One study in 1982 found that trailing tree branches accounted for only 23 per cent of channel roughness, compared to wooden piling protecting a bank, which increased the roughness by 50 per cent.[3] Secondary flows, which make a river wind around, are also insufficiently understood, and are not accounted for in Manning's formula, which is used by more traditional engineers to implement a 'play it safe' prejudice against trees, confirmed by their memories of tree-blocked bridges in the 1952 Lynmouth flood disaster.

In fact, many engineers know that the proper use of such a formula is as a broad guide, rather than an absolute master. They would probably welcome an alternative to Manning, provided that it could be simply and easily applied in practice. Scientists and geomorphologists are fast approaching such an equation,[4] especially with the availability of computers, but in 1988 the perfect solution, accepted by all engineers, is still out of reach. Environmentalists would naturally welcome a solution which is more precise in purely engineering terms, since they do not want to see rivers meandering according to a pattern laid down in obedience to some arbitrary aesthetic. If, within the gentle, but practical, restraints of preventing it from permanently going over the top, a river can be allowed to behave as it wants to, in physical terms, it is more likely to *look* right as well. Conversely, if it looks right, a river is more likely to be relatively stable in engineering terms.

A further cost constraint, affecting the shape of a new channel, was the expense of excavating shallow slopes, due to the extra quantities of earth which required removing, especially when the water-level was lowered. This led to very steep sides to the channel. The problem was compounded by a deep anxiety instilled in all engineers that any land lost to farming would detract from the whole point of land drainage, which was to maximize farming potential. Remembering the wartime food blockade and answerable in the last resort to the chairman of the land drainage committee, who was always a farmer, the traditional river engineer set about extending all areas for agriculture right up to the edge of the water-course. He did this by creating the steepest riversides which the local clays could sustain, neatly killing two birds with one stone—in this case, whole families of ducks and moorhens—by dumping the excavated river rubble and silt into nearby ponds, marshes, and meanders, which not only were convenient for swallowing up the spoil, but, in their unfilled state, were inconvenient for farming efficiency. Thus was created the now infamous 'trapezoidal channel' recommended as the basic formula for a new engineered watercourse. A trapezoid, of course, is a mathematical figure, similar to trapezium; so eloquent is the basic jargon of a fundamental clash of values: rivers as geometry.

## The Impact on Wildlife

Hardly surprisingly, local wildlife, evolved to exploit every niche and irregularity in a complex river system, did not take kindly to living in a trapezoid. Studies carried out in 1981 and 1982 on the river Ouzel in Bedfordshire showed a drastic reduction of breeding moorhens as a result of engineering work.[5] Ten years earlier, a bird census on the river Stour at Hammoon in Dorset demonstrated significant drops in the populations of sedge-warblers and reed-warblers, following a dredging scheme which also removed the nesting site of a pair of

Machines slice out a cruelly geometrical channel. The traditional approach.

kingfishers and all the potential sand-martin nest sites along the reach affected.[6] As a result of repeated heavy management, the river Lee in Essex, reported as a fine river for crowfoot in the 1930s, had become virtually devoid of such vegetation by 1981.[7] Indiscriminate tree and shrub clearance along rivers has naturally had a damaging effect on all forms of wildlife. On some upland rivers, pied flycatcher and redstart nest in riverside trees, feeding not only on insects in the tree canopy, but also on those which emerge from the aquatic margins of the river. Alder trees are a great glory of many lowland rivers, and their cone-like fruits are sought throughout the winter by flocks of redpolls and siskins. Siskins on alder, deftly balancing under each twig as they work their way from cone to cone, represent a classic example of those special ecological relationships which hallmark our river systems with their very particular sense of place. Where alders put up suckers within a river channel, nobody could reasonably defend their retention; but heavy tree clearance, in part to gain access for dredging machinery, has taken a severe toll on riverine wildlife. Studies made by the RSPB in 1978 on the river Wye in Wales showed 11 breeding species of bird with an overall density of 43 territories per 10 kilometres on a managed reach, compared with 25 species and 146 territories on an equivalent unmanaged reach. The impact on fish populations is equally critical. It is estimated that fish are provided with four times as large a food source from insects falling into a river from tree-lined banks as from bare banks.[8] Work on the upper Soar in Leicestershire in 1982 showed a reduction of 76 per cent in the mean standing crop of the fish population as a result of tree clearance.[9] Imagine, therefore, the impact upon the wildlife on the main rivers of Essex, where it is estimated that up to 70 per cent of trees were cleared between 1879 and

A siskin on alder.

1970.[10] In 1986 the South-West Water Authority was still carrying out so-called pioneer tree clearance on some of the lovely rivers of Cornwall.[11]

Learned papers and statistics only go to support the evidence of our own eyes. The Black Brook in Loughborough—black, certainly, but scarcely deserving any longer the epithet 'brook'—was given a thorough treatment in the late 1970s to reduce flood risk to an area of new housing. One side was concreted, and raised flood-banks were erected tight against its edges, without even the excuse that land would otherwise be lost to farming. Behind these banks is an apology for a children's playground which, with greater forethought, could have had the banks set round it, thereby allowing the river itself more space to meander within the playground. After approximately 8 years, no water plant or regenerating shrub has recolonized the desolate margins of the Black Brook. None the less, the local children, drawn by the fascination of running water, attempt to climb down the slippery cemented sides of this channel: they are thus in far more danger here, especially when the river is in full spate, than they are beside the more sympathetically managed reaches downstream.

It was such examples of river abuse, accelerated by the resources available from the new water authorities set up in 1974, which, by the late 1970s began to prompt a common realization that there must be a better way to manage rivers. Environmentalists and engineers began to look around for places where similar problems had been encountered and attempts had been made to come up with compromise solutions. They found part of the answer in Bavaria.

## The Bavarian Experience

In the alpine landscape between Munich and Berchtesgarten, the river Saalach forms part of the boundary between Austria and West Germany. A little further north, the Danube forms the frontier between these two countries. It was these rivers and their tributaries, such as the Vils, the Isar, and the Paar, which

Eight years after traditional river engineering, this little river looks as sterile as ever. Black Brook, Leicestershire.

were the centre of fierce fighting in the early months of 1945, as the Allies closed in on Hitler's 'Last Ditch Stand', and which in the 1960s and 1970s became the subject of a new kind of co-operative river management pioneered among the different professions working in the Bavarian State Water Authority. From the 1930s onwards, many German rivers had been subjected to extremely severe engineering. Surveys made in 1976 showed that heavy management had eliminated nearly all the tree cover from streams in the sandy plains of north-west Germany. As large-scale schemes in the 1950s and 1960s increasingly began to affect the flood-prone alpine torrents of Bavaria, where the dominant farming lobby had ensured that its rights were written into the State Constitution, there also dawned a realization that the quality of river landscapes needed defending. One reason was the importance of the countryside of south Germany for tourism; but more profoundly, the landscape of Bavaria, with its elegant Baroque churches commanding the high plateaux and the improbable Gothic castles of King Ludwig dramatically crowning the alpine ravines, has a special place in the heart of the German people, which the cascading river Sur and the winding willowy river Vils seem to encapsulate. In the 1970s environmentalists working for the Bavarian Water Authority drew up and implemented proposals to protect, and even extend, wetland habitat and tree cover, as part of practical drainage schemes planned on these rivers.[12]

Thus in Germany, and very soon after in Britain, river managers began to develop new codes of practice to ensure that, by using imagination and ingenuity, they took on a properly integrated environmental brief whenever they carried out operations on a river to reduce flooding. This new approach was actively encouraged by environmentalists, who seized upon opportunities for constructive compromise, which were a welcome relief from the conflicts which had so often dominated discussion of the countryside.

Environmentally sympathetic river management requires that ecologists work closely with farmers, machine-drivers, and engineers. We will take each group in turn and see how each has been able to contribute to this creative partnership.

### Partners in River Management

*Machine-drivers.* The drivers who operate the dredging machines do not just go home to watch David Attenborough on television at the end of a day spent mindlessly stripping out rivers. They enjoy the rivers and observe them, probably for longer and at closer quarters than anyone else. The moments when a machine-driver arrives early in the morning to find a kingfisher perched on the stationary machine bucket, patiently watching for a fish, or, as on the river Tern in Shropshire, actually taking a ride on the weed-rake, are always relished, despite much joking about catching it with the end of the jib. So is the sight of a dabchick fussing around the pools in the wake of an excavator as it dives for stirred-up morsels of food. Dabchicks are a river's equivalent of seagulls following the plough; and, best of all, there can be a bonus, when the machine scoops up an eel or a fat pike to take home for supper.

Machine-drivers are knowledgeable countrymen, often ex-gamekeepers or small farmers. It was a driver, not a professional conservationist, who showed me where the best nightingale woods in Worcestershire are. It was a driver who told me where the white ragged robin grows in the damp Warwickshire pastures which Joseph Elkington tried to drain in the reign of George III. In Leicestershire, machine-drivers have their own names for river plants, such as 'water onions' for the true bulrush, *Scirpus lacustris*. If you think of onion plants, this seems an apt description of the bottle-green tubular bulrush stems, crowned by their untidy flower tassels and swaying gently in the river eddies. In Shropshire, drivers call the butterbur which grows beside their rivers 'gypsy rhubarb'. Some also have names for particular reaches of the

Kingfishers have been known to perch on the bucket of the stationary dredger.

river, unrecognized by official maps. On the Leicestershire Soar there is a pool which local drivers call 'Dead Man's Hole', where a swimmer drowned, and one man who had spent his life working on rivers told me that he had christened a certain bridge after the small boy who spent a summer watching him build it twenty years before.

There is nothing that these men cannot do to enhance a river, given the opportunity provided by a proper environmental brief. They are quick to recognize the orange tuberous roots of meadowsweet when the plants are dormant in midwinter, in order to salvage them. The precision and craftsmanship with which drivers were previously expected to carve straight lines and obliterate habitats can now be turned to the retention and creation of a river landscape of infinite interest and variety. And how much more enjoyable it is for them and for the engineers who direct them than the old straight-ditches approach!

The dabchick follows the river dredger for stirred-up morsels.

Mighty machines can now be seen transporting water forget-me-not to newly created margins at the water's edge, the emerald-green cushions starred with their delicate baby-blue flowers trembling over the edge of the great steel scoops before being laid gently down into a new niche of carefully prepared river ooze. The impact of machinery originally designed to destroy habitat can, when properly used, be nothing short of magical. A 'Swamp Dozer', a machine so colossal that it literally makes the ground tremble, is able in a matter of days to pull back a whole field of builders' rubble or rye-grass to re-create the kind of swamp it was originally designed to destroy. In a few hours, it can hollow out a sizeable pond. The response of the natural world to such activities may be immediate. In 1984 one of a dozen such ponds was created beside the Warwickshire Leam as part of a land-drainage scheme. Before the machine had actually finished, a not-so-shy moorhen was building its nest on the island which the driver had created for it as a relatively safe refuge from such nocturnal marauders as hedgehogs, which snuffle out moorhen eggs as a special delicacy. On a river near Evesham, the machines, as part of a flood-alleviation scheme, extended a large bay and made a pool within the river bed. No sooner was the new pool created than the first swallows were dipping and swooping over it; and, best of all, by late afternoon of the first day, the bank top was crowded with bicycles thrown down in the grass by the village children while they enjoyed their new bathing-pool.

*Farmers.* It is impossible to achieve integrated river management without the agreement of landowners, who, in normal circumstances, own their riverside plots as far out as the centre of the watercourse. It is very unusual for water authorities to own land beside or within the bed of the rivers they manage.

I have sometimes been assured that opportunities for creating habitat in the countryside will always be doomed by the niggardliness of the agricultural community in relinquishing their precious acres. After 10 years of working with farmers ranging from those struggling with dairy quotas in parts of the West to the barley barons of the East Midlands, I can honestly say that the generosity and genuine interest shown by a large number of farmers in creating riverside areas for wildlife on their land still surprises me. Of course, farmers, like any other group of people, embrace all shades of opinion and responsiveness, but if two-thirds of the landowners along a river bank are as enthusiastic as I usually find them, then the future of creative river management is assured. It is no good approaching farmers as some river engineers in the past have done, with a take-it-or-leave-it 'We're doing the river, and you can have a few trees if you like.' Nor is it ever productive to fall into that cardinal error of trying to tell a landowner what is good for him on his own land. Positive, enthusiastic persuasion on behalf of the landscape by those promoting a river scheme is required; and if all else fails, behind every farmer there is frequently a farmer's wife or even a farmer's mother who will take up cudgels on behalf of the landscape beyond her kitchen door. Not that it is often necessary to go to such lengths. One leading farmer told his local water authority in 1986 that he was pleased that it was now taking proper river management seriously, since its brutal treatment of rivers in the past had given the farming community a bad name. He had a point.

With some farmers it is possible to go only as far as tree planting. Others will agree to ponds, and then go on to consider the more adventurous idea of wide river margins. Cover for game is also good cover for wildlife. It seems unnecessarily high-principled to refuse to create a pond, ostensibly so that a farmer can fire off at a passing duck (which he will probably miss anyway), when a whole ecology of herons and frogs, warblers and worms, will otherwise be the losers. One of the best examples I know of direct action against unsympathetic tree clearance under a land-drainage scheme was taken not by a conservationist, but by a farmer concerned about the resulting loss of pheasant cover along the river banks. Rightly incensed by burning pyres of riverside willows, he impounded the water authority machines in one of his fields, and took away the key. The beneficial effect upon the environmental standards of the scheme after that was remarkable.

Most satisfactory of all are those occasions when the self-evident delights of a river ensure a common purpose in defending it. I was once haggling with a farmer in the hope that he would agree to the retention of an old meander. Suddenly the blue spark of a kingfisher shot down over the water like a low firework. It sealed the bargain. 'You've won,' he said, 'we'll keep the loop open.' I sometimes take my family swimming in a Worcestershire brook which was recently given a maintenance dredge to reduce flooding, but a dredge done according to all the new principles of habitat enhancement and protection and with the full support of the relevant landowners. In the summer following the dredge, it was possible to swim down through crystal-clear water among the snakelike undulating stalks of yellow water-lilies and to touch the grips left in the soft mud by the machine, which had deftly retained sufficient stands of juicy arrowhead and bur reed to support the fluttering ink-blue damselflies. But the surest vindication of this ecological care was the fact of sometimes meeting the farmer and his family enjoying a dip in the river as well.

A farmer's involvement in the decision-making process leading to an integrated drainage scheme on his land is not only essential on practical grounds. It is also correct in landscape terms, since the individuality which stamps the English countryside with such a complex and irregular beauty is the result of generation upon generation of human decisions made on grounds of both practical expediency and individual preference. For example, some farmers happen to prefer cattle grazing down to the water's edge. In such places, cattle drinkers are put in, and shallow banks are retained. Waterside plants will then develop into what ecologists call 'poached communities', dominated by those aquatic specialists which best survive a constant trampling, or 'poaching', under the beasts' hooves, not to mention the occasional munch: celery-leaved buttercup, brooklime, and water-plantain with its finely structured candelabras of white flowers. The neighbouring farmer may well want to keep his cattle out of the river. Accordingly, a fence is erected, trees are planted, and a riverside thicket develops. In this way the eco-engineer is acting only as a guide to help the farmer leave his own individual mark on the landscape which he is creatively evolving.

The different character of every parish reflects to some degree the diverse personalities which created it, and never more so than today, when large-scale technology affords farmers the means to effect massive transformations to bequeath to succeeding generations as their personal memorials. It is easy to read the modern English landscape at a glance, picking out farm units which have been totally cleared in the pursuit of profit, as well as those which have been managed by individuals with a more balanced respect both for the necessary income and for the pleasure which their own fields can afford them.

While a river manager can act as a guide to civilized, yet practical, farming, he must still resist the pressures which come from a minority of farmers to remove over-zealously that habitat which does not worsen flooding. If a public authority is responsible for a river scheme on which a landowner refuses to co-operate in protecting the environment, that scheme can always, as a last resort, be dropped altogether.

On those occasions when drainage contractors act in accordance with an uncontrolled brief and are paid directly by the farmer, though sometimes with the help of a Ministry of Agriculture grant, it is still standard practice to clear the whole valley as part of normal operations. Down go the trees, and up go those tell-tale signs at the farm gate: 'Oak logs for sale'. It is essential that such activities by rogue farmers are controlled—by tighter legislation if necessary—in the interest, not least, of those in the agricultural community who are

Where cattle graze and trample the river's edge, a specialized plant community develops. River Blithe, Staffordshire.

committed to the creative evolution of the landscape, which they wish to pass on to their heirs. The whole community must also be ready to challenge that questionable argument which is always trotted out as a cover for unbridled exploitation: 'Since it's my land, I have a right to do anything I like with it.' Owners of historic buildings have now got used to the idea that while they can modify their homes, they do not have unlimited proprietorial rights over them. The same balanced view will need to be accepted by the owners of the historic English landscape.

*Engineers.* As far as river landscapes are concerned, the professionals who hold the key to a more balanced standard of management are the engineers, the men with the mud on their boots who face the real problem, with responsibilities both to control flooding and to take care of the environment. Taming the flood is what they are paid to do, and if flooding gets worse, they get the blame. Their problems are worsened in England and Wales by the fact that a number of our watercourses are what geographers call 'misfit rivers', small streams, compared with many of the world's great waterways, which flow along the bottoms of disproportionately wide valleys gouged out by glaciers in the Ice Age. Consequently, a minor overflowing of a bank can lead to many acres of land going under water.

Consider also the predicament of the resident engineer, the man who is responsible for seeing the job through from his isolated portacabin down some riverside track. The moment he arrives at 8.30 a.m. the more enterprising of the local farmers will come knocking at his door, asking if he wouldn't mind felling that inconvenient tree down by the river. At 9.00 the environmentalists will appear, vigilant in the defence of such riverside habitat. At 9.30 the contractor calls in, and casually reveals that since his labourer was on site before anyone else and was both unaware of the engineer's earlier instructions and in the powerful position of actually holding the chain-saw, he felled the tree in order to speed up his access to the river bank. By 11.00 a full-scale argument will be raging around the stump, and the engineer's boss will arrive, wanting to know why time is being wasted on such trivialities, and reminding him that the job had better not run over cost or behind schedule.

Under such circumstances it is remarkable that so many engineers up and down the country have given the lead in integrating environmental considerations with those of reducing floods. In 1984 the Royal Society for the Protection of Birds in conjunction with the Royal Society for Nature Conservation published a handbook which cited example after example of sympathetic river management recently instigated by engineers from Lancashire to Kent, from Northumberland to Cornwall.[13] This proved once and for all that it could be done, and that it was no longer possible to deny that treating a river as a river in the fullest sense of the word was also compatible with the effective and necessary control of flooding. Nor was the new ecological approach to engineering prohibitively expensive or impossible to achieve in terms of taking land from farmers. At last it could be said that, except in those few, but critical, cases in which adjacent wetlands would be drained away, thereby making any scheme indefensible, rivers could be managed so that they functioned as drains in times of flood and so that their wildlife diversity and landscape quality could be retained, and even enhanced, in a way that would inspire fresh generations of Constables to paint them and Kenneth Grahames to write about them. In the few years since the 1984 handbook came out, the more enthusiastic engineers have begun to convert their colleagues to working to a wider brief; and the ecological creativity of which the previously destructive machinery of land drainage has proved itself capable has begun to surprise even the most enthusiastic. In this manner, such engineers are leading the way for a multitude of other practical men whose work can have a profound effect on the countryside. They are also working within the great tradition of their profession, as defined by the charter of the Institute of Civil Engineers: 'the profession of a civil engineer, being the art of directing the great sources of power in nature for the use and convenience of man'.[14]

The concept of 'use and convenience' surely embraces a set of values which contributes to the quality of our lives, as well as the bare provisions for our survival. In the spring of 1986 a senior official from the Ministry of Agriculture addressed an audience of engineers and environmentalists in tones of genuine puzzlement with the query: 'But when you've got all this riverside habitat, what do you *do* with it?' That a man who has considerable influence on the landscape could still be asking such a question in the mid-1980s, when it is almost impossible to turn on the radio or the television without hearing discussions of the environment, indicates that the casual treatment of our rivers still continuing in some places only echoes some people's very casual attitudes towards them. Clearly the answer to his question still needs to be spelled out. What do you *do* with sunshine, or music, or children? You enjoy them, because they make life worth living. Sympathetic management of river landscapes requires no narrow utilitarian justification, although, as we saw in the previous chapter, there are, in fact, plenty of functional benefits to be gained from the ecological engineering approach. A questionnaire sent out by the Countryside Commission in the early 1980s found that people regarded a beautiful countryside as second only to safe streets in contributing to the quality of their lives.

That such priorities were held by townspeople as well as country people helps provide a simple answer to that other criticism which some cynics still level at river managers' efforts on behalf of the landscapes in their care: 'What is the point of digging out ponds and planting trees beside miles of river on private land unseen by the public?' Water-lilies and wild ducks are not validated simply by the number of people who set eyes on them. We give money to 'Venice in Peril' without feeling the need to get our money back by submitting every stone of Venice to a personal inspection. In addition, rivers, complete with their dependent ecology, are total systems, veins in the body of the English landscape.

Therefore, if the unseen reaches of a river are rich in wildlife, there is more chance that those who pause on a bridge or on a riverside footpath will have their day brightened by the sudden flash of a kingfisher, or be able to enjoy the lazy flight of a heron as it lifts slowly up over the willows.

## The Cost of Conservation

Such people have long been footing the bill for mile upon mile of habitat removal, both as part of their rates paid to the water authorities since 1974 and as part of their taxes which the Ministry of Agriculture uses to finance the major land-drainage schemes carried out by the water authorities. In Melton Mowbray stands a building originally designed to serve agricultural officials. On either side of its portals are carved the proud words 'By Farmers. For Farmers'. In itself this is a perfectly respectable sentiment; but, as applied to the agricultural industry since the War, and especially to land drainage of which that is a part, it makes me long to seize a hammer and chisel, and carve over the doorway the essential qualification 'But with the Public's Money'. That the public is also prepared to pay for the protection of river landscapes is proved by the mail-bags of supporting letters which the Severn–Trent Water Authority received in the spring of 1986 after announcing its expenditure on river habitat enhancement. General letters of approval included a number which echoed the sentiments of a couple from Rugby who wrote: 'Now we shall pay our bills twice as willingly!' Such an admission from ratepayers, as rare as the scarcest riverside species, is the stuff which senior managers in the water industry dream of.

Nor is the cost of such evidently acceptable environmental practice exactly overwhelming. In the mid-1980s, the 5 per cent of the total land-drainage budget spent on nature conservation in some parts of the country achieved astonishing results. It would be sad for the future of our rivers if finance departments were to seize on this figure as a maximum never to be exceeded; but it does indicate that our river landscapes can be given the environmental care they deserve for the kind of money which only the meanest of river managers might object to. If it is claimed that such a small extra budget for landscaping might put the cost benefit of certain schemes into jeopardy, then clearly their estimated cost was so nicely balanced against their assumed benefit as to allow for no other unforeseen over-expenditure, thereby rendering the cost benefit invalid anyway.

As will become clear, it is possible in terms of both cost and flood control to protect and extend the habitat and landscape quality of rivers as a legitimate part of the land-drainage brief. This can be achieved by walking the river and talking through the various problems with the engineer, the farmer, and an environmentalist. Whether the latter is a landscape architect or an ecologist or both, he must have in his hands an adequate ecological survey of the river in question produced, or at the very least sanctioned, by the Nature Conservancy Council and the local naturalists' trust. The creative compromise which should result needs to be recorded on a detailed drawing, which can then be followed by the water authority work-force should that be the body implementing the scheme, or else built into the tenders for any work let out to consultants or contractors. It is fundamental that all environmental proposals be planned and agreed from the very earliest stages. To cut straight ditches, fill adjacent ponds with the dredgings, and then turn up lamely and late with a few expensive saplings in the hope of concealing the devastation is yesterday's bad way of doing things. Before work starts, it is important to talk through the proposals with the digger-drivers, who will probably be able to contribute some practical suggestions as well. Watching over a scheme as it progresses is also essential. I well remember a drawing I produced for a land-drainage scheme in which I specified that a certain tree should be retained. After some time I returned to the site, only to discover that the tree had been felled and the drawing used as kindling for the bonfire!

# 7

## Creative Flow
### Rules for Good Practice in River Management

THE solution to good environmental river management lies in a number of techniques which one might call river-engineers' golden rules. These rules all comply with three fundamental principles: to leave existing habitat untouched as far as possible; where this cannot be done, to retain at least some habitat so that it can recolonize disturbed areas; and to leave such areas as physically diverse and uneven as possible, in order to speed up recolonization by wildlife. Good river management also aims at maintaining, and in many places extending, a buffer for wildlife between the actual watercourse and the adjacent land. In the countryside it is desirable to push back the cultivated land from the riverside, so that the river corridor amounts to more than just water and wheat. On most urban rivers, apart from the steel-sided drains in many city centres, where there is little room for habitat creation, the problems and potential for river managers are similar since, in such cases, playing fields and over-mown parklands generally replace intensive agriculture as a sterile habitat sweeping right down to the water's edge.

*Working from One Bank*

The most fundamental rule for environmentally balanced river engineering is that the machines carrying out the scheme work from one bank only, thereby leaving the opposite bank entirely untouched. This is frequently a sensible economy, since access for dredgers from an open, rather than a wooded, bank is bound to be easier. When the river Rea in Birmingham was brutally stripped out from both sides as part of a scheme promoted by the local council in 1982, the contractor remarked that the instructions given him to take out every single tree on either side of the stream had made the job considerably more expensive than it would have been otherwise. With some forethought and pre-planning, machines can alternate between sides in such a way as to leave unscathed those stretches which have been identified as the most valuable for wildlife. The untouched bank will provide a reserve from which waterside plants and the

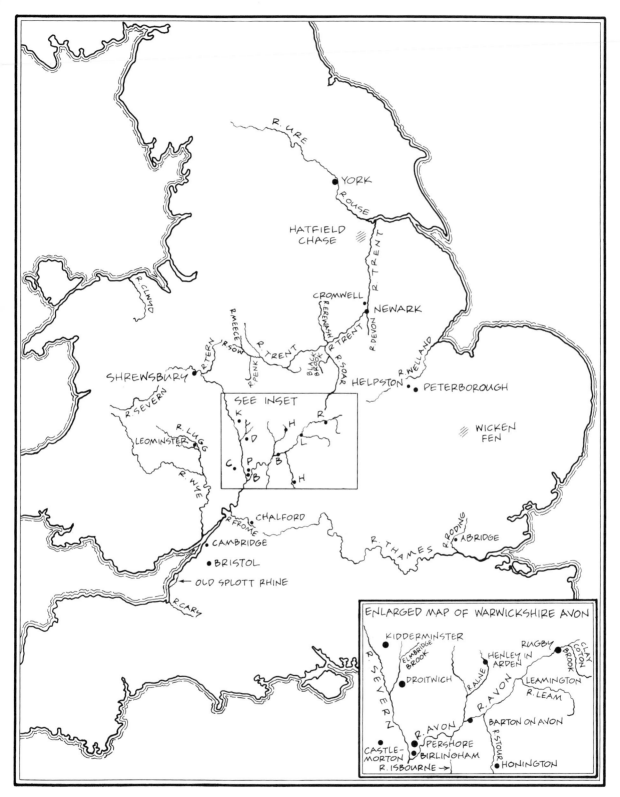

Some rivers which have benefited from sympathetic management.

herbaceous and shrubby species above them can recolonize the rest of the river-side. Another benefit from adopting this principle is that the retention of all the bends will preserve irregular corners, which farmers will accept for tree planting. A straightened cut leaves no such areas. The landscape drawing should indicate which banks the machines will be working from, together with the exact change-over points.

Old hedgerow boundaries along a river are an obvious priority for retention in a scheme which adopts the rule of working from one side only. In May, such streamside hedges in Lancashire and Derbyshire will shimmer white with the flowers of bird cherry (*Prunus padus*), whose airily descending lamb's-tails are a speciality of northern England. In May too, the older brookside hedges in Worcestershire are dense with the white blossom of wild pear, solid as clotted cream. Pears figure on Worcester's coat of arms, and pear-trees, much admired by Horace Walpole in the eighteenth century when he passed that way, are still a special feature of the Worcestershire landscape. But of course the white hedgerow flower which epitomizes, and even takes its name from, the month of May is hawthorn blossom. Those familiar may trees, which were the main constituent of enclosure hedgerows, have an especially wide distribution in some parts of England, such as the East Midlands, where they reinforce the local character of such counties as Huntingdon, Nottinghamshire, and Lincolnshire. How easy it is for unplanned clearance work along river boundaries to unstitch that sense of locality. As one Leicestershire farmer ruefully remarked in the spring of 1986 as he surveyed the stumps of some stout old thorns which had been thoughtlessly removed in a matter of minutes by the chain-saws of the local drainage men: 'They wouldn't have done that if they had only had an axe.'

A classic example of a hedge deliberately retained, rather than removed, by a land-drainage scheme lines the Whixters brook, one of many streams which flow off the Cotswold scarp and across the Vale of Berkeley in Gloucestershire into the river Severn. This hedge is dominated by massive ash and willow trees, but also contains a rich tangle of shrubs such as spindle, dogwood, and buckthorn. As ill luck would have it, at the time the brook was subject to a major scheme in the late 1970s, the farmer who owned this magnificent hedgerow said he did not mind one way or the other whether it was removed, whereas the landowner on the opposite treeless side declared the strongest opposition to any machinery gaining access to the stream via his land. The easy way out would have been to go in with the chain-saws, but the engineers concerned persisted in their attempts to persuade the reluctant farmer to let them cross his land. A meeting was held, at which the water authority solicitor and the farmer's agent stood around awkwardly in the farmyard, nervously clutching their respective files. A compromise solution was reached concerning standard compensation for disturbance, and at little extra cost to the scheme other than that of time and trouble to its originators, the trees were saved. At around the same time, engineers designing a scheme on the river Alne at Henley-in-Arden in Warwickshire retained the

You would not guess that this brook had shortly before been subjected to a land-drainage scheme. Existing trees along one side have been retained and a few more planted on the opposite bank. River Alne, Warwickshire.

trees along one bank, and also went a stage further by planting small extra plots opposite, carefully positioned so that machinery could swing in on either side in order to maintain the brook when occasion arose.

## Untidy Banks

Having agreed to work from one bank, however, there are few sites more depressing than those on which every bend and curve of the river has been faithfully followed, but on which the bank which *has* been worked has been sliced out with geometrical precision to an exact angle of $1:1\frac{1}{2}$, neatly shaped at the top with a clean sharp edge where the symmetrical slope meets the dead-level bank top. Engineers who train machine-drivers to produce such impeccably

Some of the birds and flowers to be found on an eroding river bank. Teasel and goldfinch, sand-martin and corn-cockle.

crafted neatness must be frustrated pyramid-builders; they certainly do not think very hard about the true nature of rivers. It has been, and continues to be, a slow job to teach river managers to unlearn their deeply held reverence for tidiness. It is a case of persuading them to be proud of 'doing a bad job', and of helping them to see that the important thing is that the river, while flooding less, should look as if no engineers had ever been there—the very opposite of the traditional approach. Such was the initial anxiety of engineers concerning this relaxed approach that slopes would still be shaped geometrically where they could be seen from bridges, and only once a river turned a corner out of sight was it felt safe to apply more natural treatment.

But once machine-drivers and engineers realized that they could allow themselves the enjoyable experience of deliberately creating a varied bank profile, newly engineered rivers instantly began to look better, as well as to provide a far richer habitat for the multitude of plants and animals adapted to colonizing every subtle gradation in a varied bank.

In some places, river cliffs could be retained, or even created. These are essential for kingfishers, which tunnel their nests in vertical banks, safe from marauding predators such as rats. They invariably choose a spot within 50 centimetres of the bank top, and anything from 75 centimetres to 2 metres above normal water-level. An overhanging branch or tree root will often provide a perch on which the parent kingfisher will pause before vanishing down his hole to deliver a morsel to his family. In the absence of such things, the provision of a perching, not to mention a gaffing, post can be useful for the bird, as well as provide human visitors to that reach of the river with hours of entertainment.

Another bird which nests in vertical river banks is the sand-martin, and there are few riverside sights more enjoyable than a colony of the brown and white martins which arrive from Africa each spring. They swoop and glide over the water from the river cliffs beside the Avon below Pershore Abbey, but above the river Monnow at Monmouth they have chosen a more unconventional nesting place: in drain-pipes built into the brick wall below the cattle market. Kingfishers and sand-martins will share their steep cliffs with burrowing bees and a whole number of annual and biennial plants which require the permanently eroding seed-beds which such places provide. These may include the stately common teasel, whose bristly domes of seed-heads attract goldfinches and other small birds in winter, or its less abundant relative, the small teasel, whose white cushions of flower are a local speciality of the steep banks on such Warwickshire and Worcestershire brooks as the Alne and the Isbourne in summer. A very scarce annual which germinated in the 1980s on the eroding banks of the river Waveney, which forms the boundary between Suffolk and Norfolk, is the corn-cockle, an unmistakable plant with its purple-pink flower nestling in a five-pointed star of soft furry sepals. None of these plants can survive a continuous, even-angled bank regrading, all too frequently capped with an inappropriate agricultural grass mix, the traditional engineering solution for all rivers. Where erosion is not a major problem, sympathetic engineers are now resisting the temptation to tidy up such attractive river cliffs.

## Riverside Margins

Sometimes it is desirable to retain or create much shallower slopes, and such damp, irregular banks will provide a niche for plants such as the rosy-flowered marsh woundwort, the magnificent purple-crimson candles of loosestrife, and the small subtle flowers of figwort, which always seem to be humming with wasps, including one species whose favourite prey is the figwort weevil. Such slopes should be not only uneven, but also extensive, in order to create a maximum buffer between the river and the farmed land. Perhaps the best reason of all for putting river banks well back is to create a pronounced base to them at the point where land and water meet. Such waterside margins, which should be as extensive as landowners will accept, have always been one of the commoner casualties of the traditional drainage man's approach, involving trapezoidal channels whose slopes plunge steeply into the water. Shallow margins at the water's edge are critical for the survival of many waterside plants and insects. Repeated dredging of rivers has meant that less abundant waterside plants have been washed away; and if there are no more of the same species upstream to re-colonize the silt bars which inevitably form in the fullness of time, then such plants as flowering rush and river water dropwort may never return to a river, or may finally reappear just in time to be hit by the next maintenance dredge. Re-establishing such plants by temporarily laying a scooped-out clump on one side and then dropping it back with the digger is a good idea; but there is no point in returning them to the stream if the plunging sides offer them no toe-holds.

Therefore the excavator needs to cut out a shallow shelf at the water's edge. Such shelves are usually most effective on the insides of bends, where conditions

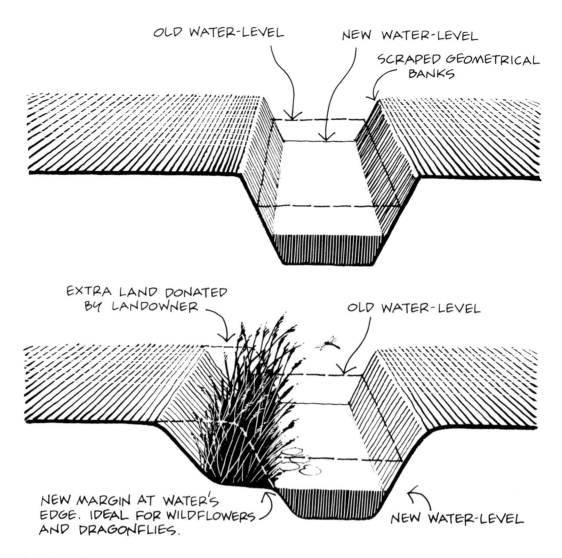

OLD WATER-LEVEL

NEW WATER-LEVEL

SCRAPED GEOMETRICAL BANKS

EXTRA LAND DONATED BY LANDOWNER

OLD WATER-LEVEL

NEW MARGIN AT WATER'S EDGE. IDEAL FOR WILDFLOWERS AND DRAGONFLIES.

NEW WATER-LEVEL

Sensitive and insensitive treatment of the river margin. *Top.* Traditional trapezoidal section. *Bottom.* A little extra land taken for a 'berm', which provides a waterside habitat for wild flowers and dragonflies.

are most favourable for their stabilization. They can grade down from just above the mean summer water-level at the bankside to half a metre below water-level further out in the stream. Since the deepened channel will have been designed to accommodate a given volume of water, the creation of a new margin will involve pushing the bank top a little further back into the farmed land. Time and again, however, farmers have proved happy to give up this small, but critical, space for wildlife. After all, they are gaining considerable drainage improvements over the rest of their acres. In the early days, engineers were anxious to spell out exactly the areas of land lost for waterside margins, in order to build them into compensation arrangements with landowners affected by a scheme. As they become more confident, however, river managers are increasingly building the retention and creation of aquatic margins into their channel design as an essential part of the package which a landowner must accept if he is to receive the benefits of flood reduction.

Through the heart of Shakespeare country in Warwickshire flows the little river Alne. It is the nearest substantial brook to Wilmcote, the home of Shakespeare's mother, and it regularly used to flood the small market town of Henley-in-Arden, which must also have been familiar to the poet, who was born in Henley Street, Stratford, a few miles to the south. Shakespeare's friend and contemporary Michael Drayton mentioned the Alne in his rhyming catalogue of rivers, which he called 'Polyolbion'.[1] In 1980 flooding in Henley was effectively reduced by a land-drainage scheme which was also designed to retain leaning willow trees and to create long, shallow, waterside margins both beside the main brook and alongside a totally new flood-relief channel. In these margins were planted flowers from a nearby threatened wetland; and the following year, even the brand new brook was bordered with a rich mix of honey-scented meadowsweet and the spikes of common spotted orchid, while, on the drier slopes of the river bank above, ox-eye daisies were established: a collection of flowers which might have been picked by the mad and drowning Ophelia in *Hamlet*:

> There is a willow grows aslant a brook
> That shows his hoar leaves in the glassy stream;
> There with fantastic garlands did she come,
> Of crow-flowers, nettles, daisies and long purples.[2]

By 'crow-flowers', Shakespeare may have meant ragged robin, whose deeply sliced candy-pink flowers resemble a bird's foot in shape; and the 'long purples' were probably wild orchids or purple loosestrife. At any rate, they would all have been flowers which were common in the district in Elizabethan times, and which will continue to flourish there if thought and care go into managing the landscapes of Warwickshire. It is not only the local flowers which persist in Shakespeare country. The name of the digger-driver who scooped out these flowery margins on the river Alne was Quiney, a family name which has long been, and remains, quite common around Stratford. In 1616 Shakespeare's daughter Judith married a local innkeeper whose name was Thomas Quiney.

At about the same time as the engineering scheme on the river Alne, another Warwickshire river, which has associations with the county's other great writer, George Eliot, was given new margins for aquatic plants by the local water authority, and with the full agreement of its engineers. A simple granite monument to George Eliot stands beside the river Anker in Nuneaton, the town near which the novelist was born and brought up. She describes Nuneaton as 'Milby' in one of her stories, after the corn mill which stood beside the river in her time and which at the turn of the century was still known as Milby mill.[3] In her novel *The Mill on the Floss* George Eliot consciously took as a model a mill beside the Trent at Gainsborough; but since she based her characters on people from

Lush margins of bur reed have been protected and extended as part of an engineering scheme. River Leam, Warwickshire, less than a year after dredging.

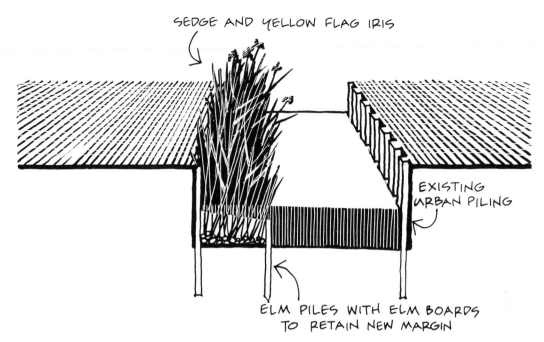

SEDGE AND YELLOW FLAG IRIS

EXISTING URBAN PILING

ELM PILES WITH ELM BOARDS
TO RETAIN NEW MARGIN

Artificial margin established in an urban steel-sided river.

her own childhood, it is hard to imagine that she did not occasionally think of the water-mill in her native town. The river at this spot, a few yards from her monument, until recently presented a sorry spectacle. The mill of George Eliot's time was burned down and replaced by another, which in turn was replaced in recent times by the present office of the Department of Health and Social Security. This building, at whose doors the dole queues regularly gather, faces the gents' lavatory across the steel-piled river Anker. The riveted bridge linking the two also carries a regular crowd of shoppers, moving between the supermarket and the main car-park. In 1980, when Nuneaton Borough Council approached the water authority to see if it could cheer up this gloomy spot, the dominant ecology of the Anker consisted of supermarket trolleys up-ended in the dark waters of the river. The engineers responsible for flood control agreed that the stream was too wide, and could therefore stand the addition of artificial margins without any worsening of the flood problems. With a small overall budget, elm posts and boards were driven into the bed of the river at the foot of the steel piling; earth was dropped in behind the boarding; British Trust for Conservation Volunteers toiled over a hot weekend to establish these newly created margins with yellow iris and sedge collected from further upstream. Within a year, gleaming stands of water plants, providing cover and food for insect life, had drawn ducks back to this previously deserted steel channel. Children were enjoying feeding the ducks from the bridge, and, miraculously, supermarket trolleys no longer seemed to be thrown in. This tiny scheme demonstrates that no river is past redemption; but it is also clear that it would have been far better to have designed the river channel properly in the first place.

*Over-widening the Channel*

Only in the centres of towns do rivers offer river engineers little space for manœuvre. Within the green corridors of washland which provide many miles of urban river with a safety valve in times of flood, as well as in the countryside, most rivers retain some marginal growth. In such places, partial dredging of the channel, combined perhaps with over-widening on the opposite side, should go hand in hand with the creation of damp margins. Ideally, one-third of a channel should be left undredged, thereby preserving a good range of existing river flora to recolonize a nicely uneven re-profiled river bed over the remaining area. On the river Leam in Warwickshire, which was dredged in 1983–4 to reduce flooding, the watercourse was extended to up to four times its original width in some places. This was achieved by pulling out wide bays in places where the river already tended to curve. In time, such overwidened reaches will silt up, but then they will provide superb damp margins for branched bur reed, yellow iris, and a local speciality: fresh green clumps of wood club-rush (*Scirpus sylvaticus*). This was achieved with the full agreement of adjacent landowners, all of whom were business-like farmers. In the spring of 1986 a pair of swans was nesting in one of the bays created by the machine-driver. At another point on the Leam it was possible to split the channel, leaving a willow, previously on the bankside, islanded in the stream.

*Pools within a River*

At other times it is desirable to retain or scoop out deep pools within a channel, in places where the river naturally tends to deepen. Such pools are important for adult fish, enjoyable for bathers, and also provide depths in which water-lilies can flourish. Sometimes, in order to keep such pools from silting up, it is necessary to impound them with a low weir on the downstream side. This contributes the additional excitement of tumbling water, and can be created quite cheaply with well-set gabions of stone-filled wire mesh. Water-lily roots float like corks, and so dredging has frequently sent them sailing off down the river, never to be seen again. On the river Blithe in Staffordshire in 1985, rivermen dropped water-lilies back into the pools, having first taken the precaution of putting the warty tubers into mesh bags weighted with a few bricks. This ensured that they would take root in the dark pike-haunted ooze of the river bed, and so grow up to brighten that river reach in future summers.

*Riffles*

Many rivers consist of an alternating sequence of pools and shallows. The wise river manager, if he wants to guide a channel so that it is relatively stable, as well as less prone to flooding, will work *with* the river, adapting the broad pattern which the stream tells him it wants to follow. In this way, while deshoaling of extensive silt bars may sometimes be needed to steady a river on its course, it is good practice to retain, together with the deeps, at least a proportion of existing shallows and riffles, always a delight of any stream, as Edward Thomas described them:

> The waters running frizzled over gravel,
> That never vanish and for ever travel.[4]

Such places are essential for spawning fish, and their rich insect life provides regular meals for birds such as common sandpipers, which arrive each spring from Africa, their gentle movements still in a dream of migration as they search for stoneflies in the low murmur of the waters.

Retention of a varied bank profile helps ensure that a river maintains this uneven bed; but if the level of the whole river is lowered, then it is often a good idea to return some of the gravel to the water. This is especially important where a shallow stream combs the long green threads of water crowfoot, whose roots thrive only in a gravel substrate. If clays or silts lie beneath shallow gravel deposits, then, after a general lowering of the bed, the gravels mixed with crowfoot roots should be dropped back into the new shallows to ensure the plants' survival. This was successfully done in schemes carried out in the early 1980s on the Warwickshire and Staffordshire Blithe. Success is also more likely if dredging is carried out in relatively slow stages, thereby allowing time for crowfoot to recolonize the disturbed areas from the less disturbed reaches. The same broad principles apply to the retention of a much scarcer plant of the river shallows, river water dropwort (*Oenanthe fluviatilis*), an aquatic member of the cow-parsley family. This plant successfully survived a scheme on the Warwickshire Leam and on the river Avon at Rugby, but a more thorough dredging in the late 1970s removed it for ever from the higher reaches of the Leicestershire Soar.

## Weed Cutting

Another constructive thing which can be done is to adapt the weed-cutting regime. Weed cutting, as opposed to dredging, is carried out on many rivers by means of weed-cutting boats, which simply mow aquatic plants from the river bed in high summer, to ensure that the combination of a clogged channel and an August thunderstorm does not result in flooding. It is frequently done on lowland rivers in response to massive summer weed growth in the channel, which in turn is a direct result of previous felling of riverside trees. Of course water-weeds thrive if they are exposed to sunlight all day and are additionally nourished by quantities of fertilizer pouring off adjacent fields with no buffer between wheat and water to absorb the superabundant nitrogen.

Weed cutting is sometimes necessary, none the less, and creates less permanent damage than dredging, since, as with mowing a lawn or a meadow, the sward—in this case growing under water—is left to regenerate. However, as with other mowing, weed cutting can degenerate into a mindless activity in which little thought is given to what is actually mown. The machine seems to drive the man, rather than the other way around. When this happens, weed cutting is not even as efficient as it might be. An example of thoughtful weed cutting was carried out on the river Cary in Somerset, by the Wessex Water Authority in 1980, in such a way as to leave unmown small stands of scarcer plants, such as narrow-leaved water-plantain and flowering rush, and concentrate on the dominant branched bur reed, which was really the cause of the problem. In this way, the slower-growing plants were given a competitive advantage over the bur reed. With thoughtful management over a sufficient period, weed-cutters should be able to reduce the number of times they mow a channel. Where channel capacity is not absolutely critical, such a careful approach has proved feasible, since the drivers of weed-cutting machinery are quite capable of recognizing one water-weed from the next. Books and identification cards can be supplied to them in the cab, and even videos are now available for the simple identification of different water-weeds, although direct tuition from an ecologist walking the banks with the driver is probably the best approach. You can be sure that the education which takes place on such occasions will be a two-way process.

It is surprising what can recover when a more sensitive approach to weed cutting is taken. The river Devon (pronounced Dee-von) flows into the Trent at Newark in Nottinghamshire. It must once have been an outstandingly pretty river, but it has been hammered by a spraying and weed-cutting regime every summer since its initial devastation by a drainage scheme carried out during the War. By 1987, when it was decided to lay off for a while, most of the growth that seemed left was a thick mucus of blanket weed, which was encouraged by the absence of both shading trees and competing water plants in the channel. A close survey revealed that just where the stream flowed past the low hummocks of a Civil War fort used by Cromwell in the siege of Newark, the river still supported living stands of flowering rush (*Butomus umbellatus*). This is one of the loveliest of our river plants. Its inconspicuous triangular leaves ensure that under normal conditions it passes unnoticed, until its spectacular globes of flower, whose petals look as if they had been cut out of flakes of frosted pink glass, emerge into the light, bringing an unexpected touch of the exotic to the waterside in July. In the river Devon, 40 years of continuous cutting had actually caused the leaves to change shape, so they were recognizable only to the practised eye of an expert employed to survey the stream. As a result of the perpetual punishment meted out to them, the plants had shrunk into themselves and had changed into a non-flowering underwater form. But they were still holding on, waiting in the bed of the river for the moment when an alteration in management would give them the opportunity to rise from the water and bloom again.

Another tributary of the Trent, the river Penk in Staffordshire, flows straight and treeless with all the graceless and undeviating monotony of the M6 motorway, which it intersects. In 1986 the engineers responsible for the Penk decided that the river was really too wide for essential summer flood flows, and that therefore they could afford to reduce weed cutting to half the width of the channel, thereby reducing the cost of maintenance. Willows could be planted at certain places along the river, paid for by the landscape budget for the maintenance work, and, best of all, a plan was devised to, as it were, ply the driver of the weed-cutting machine with drink. He was directed to mow a calculatedly weaving course down the stream, so that surviving bulrush had a chance to recover; and water-lilies were re-established in its unmown margins.

## Meanders Retained

The best local stands of water-lily available for this operation were growing in a loop of the neighbouring river Sow. They had survived there only because engineers working on the Sow some years earlier had adopted another basic technique of sensible river management: retention of cut-off meanders. Best practice generally minimizes the cutting-off of meanders, but sometimes it is necessary to speed the water on a straighter course. In such cases it is desirable not to fill in the abandoned loop, and the cost of transporting the material excavated from the new cut further afield must be allowed for in the budget. An old meander can be retained either as an isolated pool or, better still, with a through flow of water, sometimes held high by a weir. In this way extra land is made available for riverside habitat—the old loop *and* the new cut—and this presents opportunities for planting between the two. Farmers are often willing to accept proposals of this sort, especially where the meandering course of a river forms the boundary with a neighbour's land. If the neighbours don't get on, then preserving the status quo of the old boundary prevents further arguments. When meanders are filled and land swaps subsequently take place to accommodate the new river boundary, the hidden cost to the scheme of land-agents' fees in administering the ownership changes is often greater than the cost which would have been incurred by carting away the dredgings. At Clay Coton in Northamptonshire, a whole series of small meanders was retained as part of a scheme carried out in the early 1980s, thereby considerably extending the width of the river corridor. On many rivers, old mill-races provide a similar habitat to retained meander systems; and in the past, they too provided a convenient hole into which to drop dredgings. Nowadays, such activities would be regarded as bad practice, and as many water-mills are now renovated as homes or even, increasingly, restored as working water-mills, active desilting of the mill-race is welcomed by the owners.

The shape of things to come as far as meander management is concerned, however, was most convincingly demonstrated in Germany in 1982.[5] The river Wandse in Hamburg-Rahlstedt, which had long been straightened out by traditionally brutal river engineering, was of particular interest to environmentalists because it flowed through what had now been designated as a nature reserve. The officials of the Hamburg Water Authority were easily able to trace the previous wandering course of the Wandse, which showed up in adjacent fields as a damp depression colonized by kingcups, sedge, and reed. In an undertaking of co-operative detective work, the water authority, a construction firm, and a digger-driver plotted the original river route over a reach of a mile or so; and then, in a superbly imaginative gesture of creative engineering, they re-excavated the old winding river course. The straight cut was stopped off, but left unfilled, and the land trapped between the reborn graceful river and the old drain was left to regenerate into reed-beds. So popular and successful was this scheme that plans were made to continue the work in further phases. Aerial photographs of this operation, when shown to English river engineers, still tend to bring on a rather nervous reaction; but it is hard to believe that some of them won't be leading the way for such ambitious habitat re-creation in this country in the not too distant future.

On a more low-key note, one reason why such spectacular operations can be so successful is because less glamorous earlier work has been done in such a way as to preserve options for more ambitious later generations of river managers to build on. That is why not putting dredgings in old meanders, medieval fish-ponds, the hollows in ridge and furrow, flower-rich pastures besides rivers, and all the other off-loading spots previously so beloved of engineers is so important. If an engineer has remnants of wetland habitat to extend and enhance, the new landscape will generally be far richer in the long run than a brand new one which has had to start from scratch.

## Flood-relief Channels

One elegant technique for reducing flood flows while minimizing damage to the environment, and one which has long been known to river engineers, is the construction of flood-relief channels—new cuts parallel to the main stream, designed to fill with water in times of flood, but remain with simply a trickle for the rest of the year. This means that the habitat in the main channel need not be hit so hard with a dredger. The same broad principle can be extended to what are known as two-stage channels, which involve cutting away, and thereby lowering, sections of dry land beside the river, which are ideally kept grazed and will accommodate winter floods.

These flood-relief channels make excellent new habitat, especially if they have a through flow of water. One such channel at Wellesbourne in Warwickshire was planted by the British Trust for Conservation Volunteers. They used ragged robin, kingcups, and brooklime salvaged from a threatened site, and further additions of tall aquatic plants such as branched bur reed have ensured the sparkling presence of dragonflies. The soft aquatic plants bow down in winter floods, thereby satisfying the requirements of engineers and conservationists

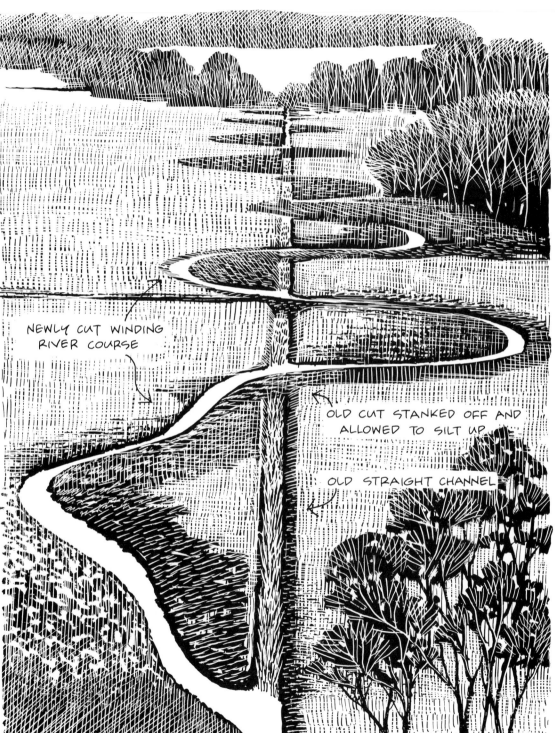

NEWLY CUT WINDING
RIVER COURSE

OLD CUT STANKED OFF AND
ALLOWED TO SILT UP

OLD STRAIGHT CHANNEL

The future face of land drainage? Putting the loops back and sealing off the straightened channel which was cut through by earlier engineers. This drawing was taken from a scheme carried out on the River Wandse in Hamburg-Rahlstedt in Germany in 1982.

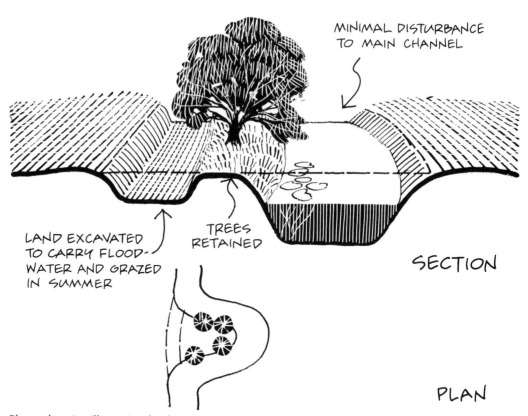

MINIMAL DISTURBANCE
TO MAIN CHANNEL

SECTION

LAND EXCAVATED
TO CARRY FLOOD-
WATER AND GRAZED
IN SUMMER

TREES
RETAINED

PLAN

Plan and section illustrating the elegant river-engineering technique of a two-stage channel.

alike. Occasional maintenance of such channels is essential, however, so that they do not develop into shrubby thickets of willow, which would trap silt and debris. Grazing or occasional return visits by the water authority are both possibilities, but best of all, the channel can be maintained by a community organization such as a parish council or a local school, perhaps led by the biology teacher. Flood-relief channels make good nature reserves to be handed over to local people if sufficient interest can be generated.

In 1980 the Thames Water Authority implemented a flood-alleviation scheme on the river Roding at Abridge in Essex. The scheme was designed to retain the river virtually undisturbed and to excavate sufficient low-level grassland beside it to carry all the extra water when the river level rose. The higher riverside land, which was previously dominated by nettles, was sliced away to allow extensive low margins of bur reed and purple loosestrife; while the retention of 95 per cent of the channel conserved a habitat for fish, insects, and water-voles. Where one farmer converted from grazing to arable after the scheme, the knock-on effect was that grazing was taken off some of the newly created low-level land. The resulting rank growth slowed down the flood-waters as they passed over this section, which in turn led them to deposit larger quantities of silt, thereby raising the lowered channels. None the less, desilting of the back channel will probably not be needed for another 30 to 40 years.

Water crowfoot in the River Blithe, Staffordshire, one year after the completion of a land-drainage scheme.

Black poplars, a rare riverside tree now being propagated and returned to the riversides. Castlemorton Common, Worcestershire.

In the same year as this classic scheme was carried out on the Roding, a similarly pioneering two-stage channel design was executed on the Herefordshire Lugg near Leominster by the Welsh Water Authority. Over a length of several miles, the meandering course of the river was retained with flood-banks which were set well back, and land was excavated on the insides of the bends. The spoil from creating these lowered areas provided a cheap source of material for building new, raised flood-banks.

Near Abbots Bromley in Staffordshire, the river Blithe once formed the ancient boundary of the Forest of Needwood. Its lower reaches remain outstandingly lovely, with massive old alder trees, under whose canopies grow stately clumps of the great northern bell-flower. The stream itself descends into the valley with a superb pool and riffle sequence, which in turn supports alternating colonies of water-lilies and water crowfoot, and is prized by fishermen for its magnificent chub. A deep pool by the bridge is a very popular bathing spot. In 1983 a scheme to reduce flooding was begun, and the principle of a two-stage channel was adopted. In this case, the secondary, dry channels were excavated in grazing land in such a way as to provide fairly straight chutes for winter floods linking the outer edges of bends on the river and retaining the alder trees between the lowered dry channels and the relatively undisturbed river itself. This was achieved by careful calculations made by the engineers with respect to the river's cross-section. As a final flourish, seed was collected from the bell-flowers, and was given to a local nurseryman to grow into plants which could be put back into the new planting areas established as part of the scheme.

## Raised Flood-banks

In the case of the river Lugg scheme, two-stage channel excavation was combined with raised flood-banks which were set well back. More commonly, raised flood-banks are simply built beside rivers to contain the flood-waters which would otherwise overflow the natural bank top. Such banks are made from earth with a clay core. They can be fatally weakened by rabbit or mole holes, or by trees growing in the bank, which, if they blow down, can create a breach. Once even quite a small hole has been made, flood-water will quickly force its way through it, causing the whole system to collapse. This is called a 'bank-blow' and men of the lowlands have always lived in terror of such blows, which explains why river engineers are anxious to maintain their raised banks free of shrubs and trees and close-mown or grazed so that moles and rabbits can be quickly detected.

None the less, traditional design of raised flood-banks, even allowing for these constraints, has been as environmentally thoughtless as have so many other aspects of river management. For a start, with the ever-compelling brief to maximize every inch of agricultural land, the banks have often been constructed tight up against the watercourse. Such banks frequently cost more to construct and subsequently to maintain, since by hugging a winding rivercourse, rather

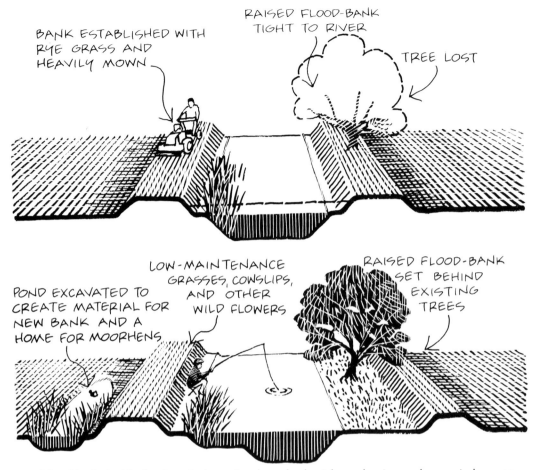

Raised flood-banks beside the river. Bad practice shows banks tight to the river and expensively mown. Good practice shows a more relaxed approach and a pond created in order to provide material to build the bank.

than cutting straight across the backs of loops and bends, they are naturally longer in extent. In addition, since rivers will always meander to some degree (unless they are encased in steel revetments), raised banks constructed too close to the river will tend in time to be eroded by the ever-wandering stream. The environmental disadvantages of setting raised banks right against a river are numerous. Existing riverside trees and hedges have often been cleared to make way for banks, when the banks could easily have been set behind them. Especially on arable land, if the banks are set well back, then a corridor for wildlife can develop. New flood-banks built beside the river Avon at Birlingham in Worcestershire were set back sufficiently for a considerable marsh to develop between the water's edge and the bank's toe, all this with the full agreement of the landowner. In winter, snipe and duck enjoy this newly available damp habitat, and in summer, flowering rush and meadowsweet thrive on the damp reserve created for them. A saving of £5,000 (at 1983 prices) was also made by engineers on this scheme by approaching the farmer to allow them to create a sizeable pond in a corner of his nearby field, thus providing material with which to construct the bank, without the high cost of importing spoil. This pond, within a season of its creation, had been colonized by a handsome margin of purple loosestrife.

In visual terms, especially as far as boat-users are concerned on navigable rivers, setting back the raised banks is very important. On the lower reaches of the Trent and Severn the raised banks, which were constructed in the 1950s and 1960s, cut off the views of those travelling up these rivers by boat as surely as if they were progressing along a motorway. If a navigated river is being embanked, a careful visual assessment should be made of all significant views, and the banks should then be positioned in such a way that they do not impede them. If the land left unprotected then proves to be so great as to break the cost benefit of the scheme, the work should be abandoned.

In 1981 the Warwickshire village of Barton near Stratford was encircled with raised banks, to protect it from flooding by the river Avon. The banks were carefully positioned behind trees and alongside, rather than on top of, existing hedgerows. Where mowing, rather than grazing, was proposed to maintain the sward, the barest minimum topsoil was spread—yet another saving on cost— and instead of the usual agricultural grass mix, a mix of low-growing fescue grasses was sown. These fine, needle-like grasses thrive better in poor topsoil, as do wild flowers, because the low, open fescue sward does not overwhelm them. A survey of the site for the scheme revealed a small horse-grazed meadow which was full of the flowers associated with permanent pasture. Since this meadow was under possible threat and was not yet considered important enough to be declared a Site of Special Scientific Interest, a small proportion of turf was dug up from the meadow, with the sanction of both the owner and the Nature Conservancy Council, and was planted in the new flood-bank. Five years later, this section of bank, situated in a caravan site, is still dominated by low-growing grasses which do not require any mowing before midsummer. This management has been ideal for encouraging a floral display which starts in April and extends until haymaking time. At Easter the bank is bright with cowslips, which have self-seeded themselves generously, and among them are the tiny bright-green spathes of adder's-tongue. As the summer days lengthen, the slopes are graced by meadow crane's-bill, yarrow, knapweed, meadow vetchling, and salad burnet, which is delicious to nibble on account of its cucumber flavour. The owner of the caravan site is proud of his flowery bank, and all the environmental care which went into the protection scheme for the village is now vindicated by an increasing trend in Barton-on-Avon in recent years: the pub is full of tourists exploring Shakespeare country. Joggers run along the tops of the flood-banks. Barns have been converted into second homes, and the local crop has begun to shift away from wheat, for which the future subsidy seems uncertain, and towards more and more caravans, which are concealed by the hedges retained by the 1981 scheme. The landscape of the Avon valley, complete with its

Raised flood-banks can be a good habitat for many wild flowers including cowslips. Barton-on-Avon, Warwickshire.

cowslips and riverside trees, whose defence required hard persuasion five years before, is now evidently part of the local economy. It is what the visitors come for.

In 1986 a more imaginative treatment of raised flood-banks was begun in a part of England in which tourism is a less certain money-earner than it is in Warwickshire, but where good environmental practice was nevertheless backed by sound economic sense. The Nottinghamshire village of Cromwell stands beside the meandering Trent. Its name, which means literally 'winding stream', 'crumb' being derived from the old English word for 'crooked', rings through English history because of the family which originated there. From this point, raised banks, last substantially rebuilt in the 1960s, extend northwards for 100 miles on either side of the river as far as its outfall into the Humber at the remote harrier-haunted reed-beds of Adlingfleet. Most of this bank is mown up to five times a summer by the water authority, at no inconsiderable cost. Nottingham-shire and Lincolnshire are predominantly arable counties, and so letting the banks to local graziers is not such an easy solution as it would be in the West. Even where farmers are prepared to bring in cattle, they are often erratic in their timing, and in some seasons decide at the last minute not to bring the beasts in at all. A late mow causes headaches for rivermen, since the lush growth by then is harder to cut and may have been concealing much-dreaded mole holes. Under such circumstances it is depressing, albeit predictable, to find that an entirely inappropriate, aggressively vigorous grass mix had been sown all the way up the Trent. For many years perennial rye-grass has been hybridized and selected by breeders to provide one of the fastest-growing and leafiest grass crops which our climate can support. Cocksfoot, which is known even to farmers as 'a good friend but a bad master', produces dominant coarse herbage in late summer if undergrazed. Because an agricultural grass mix is what drainage engineers, and even road engineers, automatically reach for, these were the grasses with which the banks had been sown. No wonder there is an expensive race for time throughout the growing season to get the mowers in.

One defence of these vigorous grasses might have been the belief that they would hold the bank most effectively in times of flood. An examination of the facts, however, proves quite otherwise. The only study carried out to date by engineers and scientists concerning appropriate grasses for flood-banks was a detailed report produced in 1976.[6] While this lacks a fully authoritative scientific standing, it does come up with a number of interesting conclusions. Low-growing, rhizomatic grasses such as fescues are best adapted to resist erosion caused by fast-flowing water. Their roots, as one might expect of grasses naturally adapted to thrive on open hillsides, are strongly drought-resistant, while at the same time so fine that they do not threaten the stability of the bank. Agricultural grasses, on the other hand, evaporate water from their lush growth more swiftly than do low fescue swards, and therefore promote drying out and subsequent cracking of the bank as well. If grazing is erratic or the growth sprints away between mowings, the tall tussocks of vigorous cocksfoot would be the first thing to encourage scour and bank erosion in a flash-flood, since its uneven surface leads to concentration of flow. Of the cases of successful erosion resistance quoted, the most spectacular example was a dam near Huddersfield, which survived overtopping, largely because, by good fortune, it supported a sward of indigenous low-growing fescues.

These short fescues thus reduce both mowing costs and flood risk. The environmental bonus from using them is that, unlike rye and cocksfoot, they also allow space for low-growing wild flowers to flourish among them, as at Barton-on-Avon. In 1986 a 3-year trial was begun on the Trentside flood-banks at Gainsborough, under the direction of specialist scientists, to see whether the agricultural sward could be eliminated with weed-killer and replaced by fescues, together with such wild flowers as hawkbit, harebell, ox-eye daisies, and bird's-foot trefoil. If these trial plots prove successful, then by a metamorphosis so gradual that its originators will probably be pushing up cowslip and cuckoo flower themselves before it is finished, the long green banks by the Trent will be transformed like immense sleeping caterpillars into what will surely be the lengthiest haymeadows in England: a hundred miles along each side of the wide river, bright with flowers and dancing with butterflies.

## Riverside Buildings

Two riverside operational sites in the Midlands indicate how the land-drainage engineer, when he has to design and build such places as part of his scheme, can adopt an approach which will reduce the expense of further maintenance and create habitat and amenity as well. At Chalford, near Stroud in Gloucestershire, the river Frome, the clearest of Cotswold streams, flows between a mill and the site which supplies drinking-water to Stroud. The engineers who reconstructed the building to house the drinking-water plant in the early 1980s had to ensure that it was not easily vandalized. But instead of constructing the traditionally vulnerable building surrounded by barbed wire, the structure itself was made windowless and as impregnable as possible, while its boundary was a stout timber fence, masked by a spiny, but beautiful, blackthorn thicket. The land beyond the operational site was then thrown open to the public, and was established as an orchard with old-fashioned fruit-trees sweeping down to the mill-pond. It is a popular spot with the locals, especially as a lunch place for workers at the mill, which traditionally produced the famous scarlet cloth of Stroud used for the liveries of the Yeomen of the Guard, but which now, more prosaically, turns out one-armed bandits. Daffodils flower in the grass beneath the fruit-trees, and these are followed by cowslips and crane's-bill until the whole sward is mown at haymaking time.

At Shallowford, beside the Staffordshire Meece, a few hundred yards from the cottage which once belonged to Isaac Walton, the famous fisherman, river engineers have constructed a little gauging station to measure flood-water levels on the stream. Such places are normally brick pillboxes perched proudly upon impeccably mown pyramids behind the statutory Stalag-style fence. Here, a little extra land was bought, to take in the whole of an awkward corner between a farm track and a railway line. The cost of this, together with the planting of guelder rose and goat willow, was soon recouped as a result of the minimal maintenance which the site requires. The existing hedges were retained as the boundary, and these have grown out into the site, especially a glorious tangle of shepherd's roses, their small white flowers cascading down to a little pond created as part of the site construction, which helped to provide spoil for the bank on which the gauging station was built.

## Ponds

It is not simply when money can be saved by providing spoil that a river engineer should be thinking of creating ponds. The presence of heavy machinery working on a river offers an opportunity for creating such habitats at a much cheaper rate than would be the case if diggers had to be brought in specially for the purpose. The loss of ponds in the English countryside since the War, accelerated by land-drainage operations, has been considerable. It is estimated that since 1890, roughly half the field ponds in south-east England have disappeared. In a typical East Midlands parish, Kimbolton in Huntingdonshire, surveys show that between 1890 and 1950, 32 per cent of the ponds were lost, and that between 1950 and 1980, a further 52 per cent were destroyed.[7] Up and down the river systems, in the course of their routine maintenance, rivermen have now begun to repair this damage.

In 1980, as part of a land-drainage scheme, a pond was created beside the Warwickshire Alne on land belonging to an even tougher professional negotiator than the average barley baron: a building developer who gave his support to the project. This pond provides welcome interest for the occupants of a new housing estate built up around it; and within a season, the warm air above its still, peat-dark surface was alive with the darting movements of swarming damselflies, *Enallagma cyathigerum*, resembling light-blue needles; while king of the pond was undoubtedly a broad-bodied chaser dragonfly, *Libellula depressa*, which took up its position on a tall reed stem, dashing out with regular ferocity through the hot afternoons in pursuit of intruding males or visiting females. Planted on the pond margins was great spearwort, our largest buttercup, which was growing on another wetland in the Midlands, one which was doomed as a landfill site for rubble created by alterations to the M6 motor-way. Five years later, a deal was agreed between the owner of the wetland, who had long before obtained planning permission for dumping, and the water authority, which owned adjacent land of less environmental interest, which it offered him for dumping in order to save a still surviving two-thirds of the original wetland. By this time, the spearwort had long vanished beneath road cones and mounds of broken tarmac, but it was now possible to plunder succulent clumps from the Warwickshire pond, and so return this scarce plant to what sur-vived of its original home. The thin green line had been held.

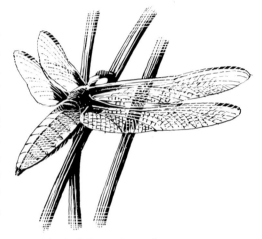

Broad-bodied chaser dragonfly.

One of the largest of many ponds created as part of a flood-alleviation scheme on the Warwickshire Leam in 1984 was made on part of a two-acre riverside plot known as the Hog's Piece. The farmer who owned it—no laggard, but a progressive arable man—gave the whole field for habitat creation, making a conscious decision to abandon the underdrainage which he had put in the previous year. It was a fine corner to choose for such enhancement. A water-mill, demolished at the turn of the century, had occupied the site since before Domesday, and there still survived a small mill-pond, a mill-race lined with old and unusually large field maples, and an indefinable atmosphere, which perhaps has something to do with long and ancient occupation. A huge pollard willow, whose massive bole, infinitely creased with age, harbours blown snow, which cleaves to its interstices in midwinter, and the young growth of festooning ivy in spring, leans out over the little pond which the diggers proceeded to treble in size. As the blades pulled back the margins of the mill-pond, an expert from the Museum Service was able to identify broken splinters of medieval glass, resembling glossy black toffee, among the sherds of blue and white willow-pattern porcelain at the willow's foot. The following May a toad, for which the newly deepened pool had provided a perfect home (toads require deeper water than frogs), was belching out its love-song, invisible among the further margins of the mill-pond. In the adjacent open field, a new and even larger pond was dug, with a pipe feed linking it to the river, to

An ancient pollard given a new lease of life as part of a river-engineering scheme. The Hog's Piece, River Leam, Warwickshire.

provide a freshener at times of high water. Dragonflies now hawk and cruise after their prey over its clear water, soaring occasionally in exuberant flight right over the willow tops. Perhaps, one day the new dragonfly populations of this pond will tempt an even more glamorous predator known to roam from the woods in this part of Warwickshire: the hobby hawk. Hobbies are especially partial to dragonflies, and observers have watched them hunting near ponds, seizing dragonflies with their claws, and rising upwards with their glittering meal, as the insect's uneaten wings flutter gently back down to the water.

### Artificial Habitats

The most important thing which a land-drainage engineer can do for the birds on the rivers in his care is to protect and extend natural habitats so as to provide them with shelter and food. There are smaller, but on the face of it, more immediate, things which can also be done, however. Old brick or masonry walls frequently provide chinks and holes for birds such as dippers and wagtails to nest in. When the time comes to rebuild them, concrete and steel or an impeccably built brick wall afford no such lodging places, and so nesting boxes can be built into new weirs or bridge abutments, or, best of all, strategic gaps can be left in the new structures. No less than eight nesting boxes were slotted in beneath the girders of a farm-access bridge constructed by the Welsh Water Authority near Leominster, and, astonishingly, the majority were occupied in the first season. None the less, while such commendable features are well worth constructing, they are no substitute for a properly thought-out, environmentally integrated scheme. On the river Blackwater in Northern Ireland, as part of major engineering works carried out on the river and its tributaries in 1986, a huge new bridge was constructed at Caledon. Standing before the colossal slab wall of the new structure, which itself was dwarfed to Lilliputian tininess by the massive scale of both the surrounding machinery and the total operation, which was shaking the entire geography, let alone the ecology of the river system, out of all recognition, you could just see, by straining your eyes hard, set like a pin-hole in the great cliff of reinforced concrete, a lovingly constructed dipper box: a room with a view, indeed!

A hobby catching dragonflies.

Similarly, artificial otter holts consisting of rock piles covered over with earth, are worthy in so far as they offer immediate cover for otters where nothing else exists.[8] But as a long-term alternative to the retention or, where necessary, re-creation of total river habitat, they are surely distasteful. To accept the idea that an open drain, with the addition of a few man-made devices to provide homes for its unfortunate inhabitants, can ever be an acceptable substitute for a proper river is to dodge the real opportunities for integrated river management. On the Somerset Levels, at a time when the whole wetland system was under threat, a much-photographed otter holt was constructed in a totally open field beside a sad, treeless river course. Three concrete pipes led uninvitingly to a cairn of brick and stone. It is hard to imagine any self-respecting otter creeping into such a squalid refuge; but even if it did, this construction could hardly be called a habitat. A proper otter holt is a tree, which provides a home for numerous other creatures besides otters, as well as gracing the river scene. Of course, ancient trees cannot be created overnight, but scrub of bramble and blackthorn, in which otters lie up in the day, can be established quickly. Generous land-take, combined with a long-term tree planting strategy, should always be part of river operations if we are to pretend that we are taking wildlife seriously. Similarly, the decline of our river landscapes has become a serious matter, if, as reported in *The Times* in November 1985, fisheries officers had to resort to creating an artificial reef of 300 tyres on one of Lincolnshire's prime coarse-fishing rivers, to provide sanctuary for roach and bream.

*Bank Reinforcement*

Seldom has disregard of river landscapes been more evident than in the case of the materials used by traditional river engineers to shore up the banks. In the 1980s, for example, car tyres have been seriously recommended by a leading engineering consultancy to the Construction Industry Research and Information Association for bank protection. What a way to grace a river bank, however efficient it might be. Another surprisingly common traditional method of binding river banks has been to use old cars. These too are listed in a matter-of-fact manner as one technique for reinforcing a bank, this time in an American standard work produced in 1977 by the United States army engineers, into whose tender care are entrusted the glorious rivers of America. (Note the mismatch of profession to task in hand.) Having suggested that the cars be roped together, bumper to bumper, along a bank, this publication proceeds: 'Large-scale use of automobile bodies as bank protection devices would improve the landscape aesthetically by eliminating some of the junkyards scattered along the nation's highways, but [it adds thoughtfully] would make the banks of streams unsightly and inaccessible.'[9]

Such methods are not confined to the United States. In 1970 the *Sunday Times* ran a feature entitled 'Farmer Tames a River with 46 Cars'.[10] A large photograph showed a Welsh farmer admiring his handiwork, with his faithful collie dog at his heel, but the most touching thing about it was the way in which he had planted sapling trees in the front seats of the half-submerged Fords and Austins. Cars are still used by farmers occasionally to shore up a river bank, in some cases with the tacit encouragement of water authorities, whose previous removal of proper bank-stabilizing tree cover led to the massive erosion which these farmers are now trying to combat so desperately.

Another more universal, and probably more obtrusive, because longer lasting, method of reinforcing river banks is to cover them with stone on to which white cement is then poured or sprayed. This finishing touch is often insisted upon to discourage local youths from lobbing loose rocks into the river. In fact it is generally the final stage in the downgrading of a riverine environment, to which the only reaction which can be expected is vandalism. Paving slabs are sometimes joined together as a bank reinforcement, occasionally with the extra flair of adopting an alternating pattern of raspberry-pink and peppermint-green tinted ones. A good example of the street-paving approach to a remote rural river, implemented in the mid-1980s, can be admired by all the motorists on the M54 as they cross the river Penk in south Staffordshire. Numerous proprietary brands of cellular concrete blocks and artificial fibre are also available to river engineers.

These at least have the advantage that they are designed to include pockets into which earth can be put, and in which natural vegetation can thus regenerate. Their disadvantage is that if they are to act as a stable reinforcement, they must be laid very precisely on the bank, which inevitably dictates an extremely geometrical angle to the slope, however green it may get.

Nobody can deny that there are occasions on which the careful guiding of a river along its course requires some bank reinforcement. However, there are plenty of sensible materials to hand for the environmentally aware river engineer. Stone from local quarries, which is often quite cheap, as well as appropriate in that it reinforces the local sense of place, has often been used. This is especially successful if the machine strips some of the soil from the existing bank, since it will be full of indigenous roots and seeds and can then be respread over the stone. The operation is done in no time, and bank recovery seems almost as instant. On the river Meece at Shallowford in Staffordshire, the bank was laid with stone and then spread with soil in this way, and within six weeks the living plant material had knitted over the stone, providing a resilient, yet attractively appropriate, living cover which included many grasses, dog daisies, figwort, and at the bank's moist toe, the juicy leaves of brooklime and water forget-me-not. This saved the cost of importing cement; was far more resistant to damage by vandals, who could no longer see the rocks to vandalize them; and was especially resistant to erosion by the river, since the tight matrix of moisture-searching roots bound the interstices of soil amongst the stones, while the native, low-growing sward, further stunted by the poor rubbly spoil in which it was growing, produced a flowery, but compact, surface to the bank.

One of the very best materials for binding river banks is ordinary English blackberry. Continual scraping and scouring in the name of drainage has reduced even this familiar feature of river banks, as John Clare noted in his poem, 'The Flitting' as early as the nineteenth century:

> The stream it is a naked stream
> Where we on Sundays used to ramble
> The sky hangs o'er a broken dream
> The brambles dwindled to a bramble.[11]

Brambles are yet another part of that local variety which creates such a strong sense of place in the English countryside. Over 2,000 microspecies of our native blackberry, *Rubus fruticosus*, have been identified by botanists. Their flowers range from white to deep pink, their leaves from glossy to felted, and their briars from the totally thornless to the ferocious, and from cream and green to the deepest shades of mulberry-purple. In his 1972 *Flora of Staffordshire*, E. S. Edees lists sixty-four varieties, some of which are woodland specialists associated with particular valleys, others of which are confined to sandy heaths, and a number of which remain the speciality of one particular parish in a county.[12] The leaves are one of the favourite foods of the caterpillars of the green hair-streak butterfly, as is indicated by its Latin name *Callophrys rubi*, so a bramble patch by the river may increase the chance of seeing this little butterfly flit by on its nut-brown and apple-green wings. On the river Erewash in Nottinghamshire in 1985, instead of stripping off and burning all the brambles from the river bank, as was previous practice, river engineers put the plants on one side and then set them back in the newly graded bank. On the Worcestershire Isbourne, which was regraded in 1982, it is interesting to see that where brambles were deliberately retained on the bank, it is holding well; whereas the patches in between, which were clear of brambles, have successively slipped and had to be repeatedly lined with stone.

Another good bank stabilizer, which also creates an attractive habitat, is the common reed. Clumps can be transplanted quite easily, provided that they are lifted and then planted in the dry and, above all, in the growing season, when they are less likely to rot from fungal infection. Reeds will colonize right up to the dry tops of banks, as can be seen on parts of the M5 and M6 motorway embankments, and their splendidly plumed stands provide nesting space for sedge-warblers and reed-warblers. There is a fairly long-standing tradition of establishing reed to trap silt and reduce scour in estuarine regions, as was done on the tidal reaches of the Ouse by the Yorkshire Water Authority in 1975; but

A green hair-streak butterfly on bramble.

the possibilities of using reed as a bank stabilizer on rivers inland, as practised on the Warwickshire Leam in 1985, are also considerable. A tough aquatic plant, confined more to the water's edge than reed, is the greater pond-sedge *Carex riparia*, whose handsome blue-green blades fringe the margins of the Lee Navigation, where they were successfully used to stabilize the bank by the Thames Water Authority in 1978.

Perhaps the most spectacular green alternative to wire mesh and welded steel for protecting river banks against erosion is a living woven wall of willow. Sharpened willow stakes can be driven straight into the toe of an eroding bank, and smaller, more pliant withy stems can then be laced horizontally between these vertical supports. This was done on the fast-flowing Staffordshire Meece in 1981, and by early summer the willows were green and growing. Bundles of willow brash laid, rather than woven, between upright stakes constitute another variant of this technique, sometimes known as 'kiddling' or 'faggoting'. It was carried out by the Yorkshire Water Authority on the river Ure in 1981. The use of large growing willows within a bank is advisable only on wide rivers, and it is as well to protect the new growth from grazing livestock. On the river Clwyd in North Wales, a river engineer came up with a particularly neat solution to a problem of bank erosion in 1977. A willow had fallen into the stream, and as a result, the river, nudged out of its regime, was eroding the adjacent bank. The engineer removed the offending tree, and chopped it into logs with which he filled mesh baskets, which he then set into the newly eroding bank. These grew immediately, efficiently holding the bank against erosion. Returning ten years later, the engineer was able to admire a riverside grove of 20-foot willows still attractively solving the problem which their parent tree had created.

## Trees

This takes us to the simplest and most obvious of bank reinforcements: trees. 'Can't you do something about it?' exclaimed an exasperated Welsh farmer to me, Canute-like, as he watched the swirling waters of the Severn tearing away at his riverside acres and the neatly grazed turves of good grass toppling down the cliff and vanishing with a plop into the all-consuming flood as we spoke. The answer had to be that it was too late. The removal of mature trees holding the bank had inevitably speeded up the erratic course of the river, which all the steel and iron in the world could no longer contain. The only possibility was to give over even larger areas of land to planting, well back from the river, in the hope that in time this might curb the Severn's course. Such is the innocence of twentieth-century man that he thinks that even something as elemental as a major river can be tamed instantly by some convenient new technology available on request through a phone call to the man from the water board. At the other end of the country, where wide lowland rivers seem to the untutored eye to flow more lazily, modern measurements prove that they in fact rush as urgently as ever towards the sea. In addition to holding the bank, trees reduce sediment loss from bank to water, and shade out the summer crop of nutrient-fed weed. Trees provide far more than just prettification of the riverside; they are a fundamental tool of those who would attempt to tame the flood. Yet, on the whole, they are still planted by engineers as an afterthought and out of the landscape budget.

Every environmentalist dreads the way in which tree planting is used as a panacea and a cosmetic to cover up river mismanagement. A good example of thoughtful management of existing riverside trees is the pollarding of willows. Once willows have been polled, above the browse-level of cattle, they require similar treatment every 7–10 years, which generally works out at the same interval as the necessary maintenance operations on a river. It is reasonable for a river engineer to also carry out this treatment on pollards which are a little way back from the river bank, such as those frequently found on old mill-races parallel to the stream. At Hinton-on-the-Green in Worcestershire some ancient pollards so stout that a man could lie feet extended across their crowns were in danger of splitting apart and collapsing through neglect when a river scheme came through in 1983. They were given a short back and sides for the then current price of £12 a pollard, which immediately revitalized them, as well as providing poles to be hammered into the river bank as a free, fast-growing, and entirely appropriate source of new trees. This planting of stout sections of willow pole is an especially neat way of establishing new trees and retaining local strains of willow; ideally it is done in winter, and new poles should be protected from grazing livestock until they get going. The ease with which willow poles establish themselves is apparent from one riverside pub garden, where the publican banged in willow stakes as a support for fairy lights, and now has an entire grove of silvery trees.

The most expensive element in new riverside planting is fencing, which is essential on grazed land. Costs can be reduced in those schemes in which an automatic part of reinstatement is to establish fencing along the tops of the new steep banks to prevent animals slipping down them, or indeed wading across the river on to a neighbour's land. With pre-planning, this fencing can be cut across spits or bayed out to allow planting on the river side of the fence. When designing tree plots, allowance should be made for the grazing reach of animals over the top of the fence. On arable land, planting costs are substantially less because there is no need for fencing.

Farmers and other landowners will accept trees on their land for a great many reasons. Arable farmers are often unable to get their large machines into awkward corners, such as bends in a river, and such fallow patches, unless planted, are a permanent source of dock and thistle, whose seeds blow into the crop. Grazing farmers, on the other hand, sometimes complain when they round up their stock, that the animals may stampede and even damage themselves against fencing in awkward corners, which are better blocked off and planted with trees. For many farmers, cover for pheasants and foxes may be almost the most important single consideration. One landowner on the Shropshire Wealdmoors in 1986 saw the land-drainage scheme on his land as principally a way of designing a more exciting shoot, with maximization of agricultural production very much in second place. Farmers also regularly stand to make a fair bit of money from the fishing they let out on their rivers, and here again, retention and planting of new trees is an important practical consideration. Where trees reinforce boundaries between landowers, it is well known that good hedges make good neighbours. Furthermore, few farmers can resist a free offer, and that includes trees from the local water authority. I well remember one landowner who held out against an offer of tree planting until he saw his neighbours along the river bank receiving the benefit of new copses. In no time at all, the telephone rang with the slightly sheepish request that he would like some after all. Plenty of farmers are glad of trees because they provide shelter and improve the view from their farmhouses, while the more politically aware leaders of the farming community realize that as grant aid for their enterprises is increasingly threatened by the bad image of agri-business farming, it is a wise expedient to be seen to be planting trees.

Sizeable, strategically placed plots of trees beside a river, compared to thin lines of trees standing sentinel along a bank top, are often better from the point of view of habitat, as well as the pattern of practical farm use and access for future dredging. Corner plots where a tree may already exist ensure a nucleus of habitat, which will swiftly enrich a new plantation. A balanced degree of planting fulfils the needs of the engineer, who requires space for future access to the river, and the conservationist, who finds that a completely shaded river supports fewer aquatic plants and insects. If there is a choice, it is preferable to plant trees on a northern river bank, in order to allow a reasonable amount of light to reach the water. Trees and shrubs planted to overhang the confluence of feeder streams with a river are particularly valuable, enabling birds and animals to travel up the corridor undisturbed. However insignificant a feeder ditch to a main stream may seem, it can be a watery artery worth enhancing for certain forms of wildlife, as was demonstrated by the case of a four-foot eel which had slipped up a boiler-house pipe in Thurcroft colliery in south Yorkshire. Fattening up for years in its tailor-made home, it finally blocked the system, until, as reported in the *Doncaster Echo* in 1985, it was discovered by a miner, who despatched it and took it home for breakfast.

From the moment a tree plot is fenced off and planted, it is surprising what a refuge it can form for wildlife. When saplings were planted in a previously cultivated field beside the Warwickshire Leam, an immediate crop of cheerful white flowers of Jack-by-the-hedge came up. As the food plant of caterpillars of the orange-tip, it ensured a positive plague of those delightful butterflies. The establishment of native shrubs such as guelder rose and hazel among the trees will ensure a good scrub layer for nesting. Blackthorn will sucker to form a dense thicket if protected from grazing and the omniverous plough. Such damp riverside scrub is ideal for the melodious blackcap and the nightingale.

Although willow and alder are always worth planting as part of a land-drainage scheme, their ability to regenerate easily has ensured that they are often the only riverside trees to survive previous intensive management, and so a tree-planting scheme offers an opportunity to return less resilient trees such as ash and oak to the river. One of our scarcest native trees is a particular speciality of the river bank. The English black poplar, *Populus nigra* var. *betulifolia*, is a distinctive adornment to any stream, with its crusty bark, its gnarled roots gripping the bank, and a descending rush of silvery boughs bristling with crimson flower cones in April. To reproduce themselves, black poplars require closely adjacent male and female plants, whose fertile seed germinates successfully only in fresh riverside mud where there is no competition from other seedlings. With such fastidious sexual habits, and further inhibited by the activities of generations of rivermen hewing, felling, and dredging, it is scarcely surprising that known specimens of native black poplar were reduced to less than a thousand in the country by the early 1970s, and that every one of them was condemned to permanent celibacy through the absence of a convenient mate. Since then a

The unmistakable crusty bark of a black poplar.

campaign to plant black poplars has begun to get under way, and in 1986, the Severn–Trent Water Authority organized the propagation of specimens existing in its Severn catchment at Castlemorton Common near Malvern, to reintroduce them to their long-vanished but ancient haunts within the catchment of the Trent.

On many upland rivers, such as the higher Ribble and the rivers of the Peak District and of Wales, a campaign is needed not so much to plant trees, as to fence them off. The trees are there, but nibbling sheep prevent new regeneration, and perform a permanent pruning operation on the landward side of the tree cover, forcing all lateral growth even further out across the stream. This can increase the anxieties of engineers, who may finally be tempted to move in with chain-saws; or it may lead to the collapse of a one-sided tree, thereby creating a blockage, as well as causing bank erosion by the hole it leaves.

## An Upland River Landscape

Since rivers constitute total, many-fingered systems, reaching up from the lowlands into the remotest hills, tree cover along their banks is especially important as a corridor for wildlife. Otters, which make their holts in cavernous hollows where the roots of ash and sycamore overhang a river, were by the late 1970s beginning to expand their populations in the head-waters of the rivers Severn and Vyrnwy, as they slowly recovered from the poisonous effects of dieldrin, no longer available as a sheep-dip. However, as the otters began to migrate down from the Welsh hills, there was nowhere for them to go, since, especially at the confluence of the Severn and the Vyrnwy, there were only scoured miles of river bank devoid of even bramble or blackthorn. The natural link between the two otter populations of the Severn and the Vyrnwy had been totally denuded of trees, as had an extensive intervening reach between the well-treed Welsh reaches and the wooded banks of the Severn much further downstream. A few otters might hurry through this inhospitable gulf, but they would hardly pause to breed and so build up a population there. In 1984 the local water authority began a long-term programme of planting this empty stretch of waterway at modest annual expense, following a detailed survey of otter populations by naturalists. A mathematical model to predict flood flows is being prepared in connection with this operation, to ensure that trees do not worsen the situation in more flood-prone areas; and landowners have proved very responsive in donating land. In fact, the policy of fenced planting is very much in harmony with farming aims, since many of the local sheep-farmers find that their livestock can fall down the steep river banks, and that lambs are sometimes carried away by the fast-rising Severn. Totally continuous planting of the river is no more immediately essential than it is financially possible. Provided that there are reasonably regular stands of trees, rather like beads on a necklace, otters can make their way from one riverside copse to the next. With time, the gaps in the necklace can be planted.

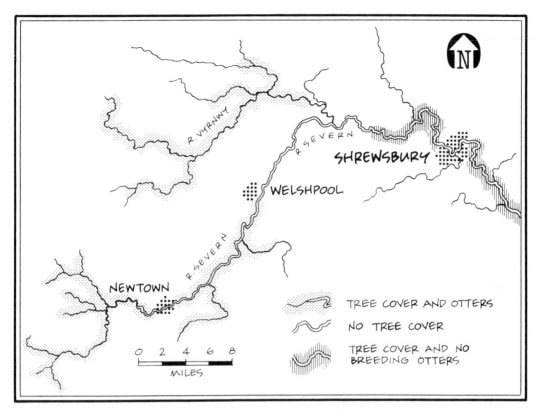

This map shows both the well-wooded and the denuded reaches of the Rivers Severn and Vyrnwy. A long-term programme of planting is under way to fill some of the gaps.

The beauty of this project is that the appeal of furry animals, which goes down so well with ratepayers—public relations departments know full well that an otter is second only to Princess Diana in the publicity stakes—combines with an upgrading of the total riverine habitat for all the less glamorous creatures which have no defenders. After seeing nothing but lowland rivers, visiting the upper reaches of the Severn and its tributaries, whose mature fringes of ash and oak have survived the riverman's axe, which has so devastated the lower reaches of the same system, is like first exposure to a symphony, having previously been accustomed to just a handful of notes. The oak and the ash let down their great boughs of green, sprouting with polypody fern and swarming with the juicy arteries of summer leaf, over the river rocks and the moving water. A dipper bobs and vanishes. There is no reason, ecologically speaking, why such sheer quality cannot be extended down the river within the lifetimes of our children, provided a persistent commitment to this surely attainable ideal is sustained by those responsible for managing the river. The return of the otter to the sad middle reaches of the Severn will then prove to be the springboard for the total resurrection of one of our finest river landscapes.

*Top*. The quality of the river habitat in the upper Severn system ensures that these rivers remain an important centre for our otter populations. *Bottom*. Further downstream the stripped-out gulf of the Severn offers little permanent cover for otters.

*A Lowland River Landscape*

The potential of river-enhancement works for breathing life back into whole aquatic systems is illustrated by another ambitious project begun at the same time as the Severn planting, but in a very different part of the country. It involves nothing less than the rebuilding of an entire fen landscape. East of Doncaster, extending dead level from the shallow rise known as the Isle of Axholme, is some of the richest farmland in England. This still remains recognizable as the landscape created in the seventeenth century by Vermuyden in his first, highly controversial blueprint for the reclamation of the entire Cambridgeshire fens. Rust-red pantiled barns, sometimes showing Dutch influence, stand moored in what seems like a boundless sea of wheat and barley. There is a sense of infinite space and silence, as if the rippling oceans of grain are just an extension of the huge emptiness of the North Sea, whose waves beat against the shore, actually above the level of this land, a few miles to the north-east. But the dusty, tawny impression of this great granary at harvest time is misleading. It is true water-land. The straight roads, bounded on each side by dykes, are ruckled and uneven due to peat shrinkage; and if you sit on the dykeside, you can feel the whole fenland shake with every passing vehicle. Intersecting every parcel of this not so solid earth are 150 miles of publicly maintained channel, with as many farmers' drains feeding into them. Seeping secretly between the standing corn, the smaller ditches empty into an ever-larger and faster-flowing web of waterways, until they converge upon three mighty rivers, flowing side by side like liquid tramways towards their confluence with the Trent. Through this labyrinth of dykes, water is pushed away with a never-relaxing urgency by pumps housed in flamboyant Victorian engine-houses built of red, blue, and cream brick, complete with stone gargoyles of water-gods in which the swallows nest. The engineer in charge of this system is a worried man, as well he might be. When rain patters on his roof at night, he gets up and walks the floor, thinking that perhaps he should lower the water-level from the engine-house controls. The waters never rest, and the fall in the dykes is a mere foot over one mile.

Where agriculture is so intensive and drainage so critical that a single clump of water-weed in a main dyke can snag the entire system, one might imagine that any space for habitat enhancement as part of the land-drainage brief is simply out of the question. But this has not proved to be the case. In 1985 the water authority responsible for maintaining the dykes and banks of this fen system began to discuss with farmers and on-site engineers the possibilities for rethinking existing management in such a way that, while its functional purpose would remain the absolute priority, the linear extended fen, still surviving in all those miles of ditches, might have a little more room to flourish. The results surprised even the most optimistic conservationists.

In terms of a sense of place, let alone of practicality, the idea of unstraightening the geometrical matrix of channels in this particular landscape would be as

Rust-red pantiled barns stand moored in what seems like a boundless sea of wheat and barley. Hatfield Chase, South Yorkshire.

inappropriate as trying to carve serpentine loops into the formal grid of a garden like Versailles (whose landscape style, incidentally, is inextricably bound up with the seventeenth-century drainage projects which were so popular at the time it was first laid out). What *could* be done was to widen the dykes where unfarmable corners of land were trapped between water and roads or other structures. Such lagoons would then provide a refuge for scarce surviving aquatics such as the specialized and spectacularly beautiful ditch crowfoots, which in the normal course of management had been sprayed out every spring in order to keep the ditches clear. Some other, less critical dykes did not, as it turned out, require vigorous bankside mowing, which had been done simply out of custom, not thought. One of these channels which, as well as bordering a quiet road much used by cyclists, had the extra scenic advantage of culminating in a magnificent engine-house, burst into flower as soon as the mowers were taken off, in what amounted to a mile-long herbaceous border. Comfrey and cow-parsley began the season, and by midsummer were followed by an extravagant display of sorrel, speedwell, bird's-foot trefoil, knapweed, and orchis. Pride of place among this haze of flowers was taken by large stands of the scarce meadow rue, smothered with creamy panicles of flower, and supporting fluttering colonies of its own specific meadow rue moth. A single, late-summer cut—far cheaper than the continual mowing done previously—satisfied the requirements of ecology and drainage alike.

On another dykeside, close inspection revealed that the monthly mow was religiously removing new growth from the living stumps of oak, ash, buckthorn, and wild rose, which were the determined survivors from the tree cover of the ancient fen. Since they were all between the road and the dyke, the farmer who owned them was perfectly happy to see the trees grow, and pointed out, with evident enjoyment, that birds moved in to nest in the oak thickets within the first season of their new-found freedom. The drainage men were happy too, since

Trees rear up above the level landscape of the Fens.

the tree roots patently held the bank, which in other places was bare and slipping as a result of the continual scrape it had received from the mowers; meanwhile, the shade cast by the oak trees over the ditch naturally reduced the need to maintain the aggressive growth of water-weed in the channel. These benefits, not to mention the safety element which the tree trunks provide to night-time motorists, since they pick out the positions of perilous adjacent ditches, provide good practical justifications for a grid of dykeside planting, which is exactly what is needed to lift this region to a dimension of real quality. In these great levels, a single tree rears above you like a cathedral; and relatively few lines of willows, marching like processions of silver smoke clouds beneath the immense cumulus of the fenland skies, seem to bring out the distinctive character of what must be one of the most individual landscapes in England. In 1986, 8 miles of trees were established along the ditchsides and a survey was carried out on all the ditches in the fenland around the Isle of Axholme, as a basis for a total remanagement of that landscape already begun the previous year by the water authority, with the vigorous support of both farmers and drainage engineers.

In the winter of 1986 another, even more startling initiative, conceived by the land-drainage engineers of the Anglian Water Authority, was implemented in the heart of the greatest level of all, the Cambridgeshire fens. Consciously exercising their environmental brief as river managers, those responsible for drainage set about reversing its usual effects, so as to ensure that the water-level rose. With symbolic significance, the precedent for this natural development of the river engineer's role was set in Wicken Fen, the very same landscape of sedge, reed, and willow over which the conflicting testimonies of Bloom and Ennion had raged 40 years before. By this time, much of Bloom's wartime drainage had been abandoned, and the land had been incorporated into the main reserve of Wicken Fen; but despite this apparent triumph of wetland habitat over agriculture, a more insidious threat to the entire system was presented by the sponge-like effects of the surrounding drained land in which the wet fen lies like a tiny island in reverse. Because of the inevitable effects of peat wastage on the agricultural land surrounding Wicken Fen, the sedge fen was recorded even in Bloom's time as being 6 feet higher than the adjacent farmland. Attempts at pumping water on to this raised pudding of peat had been abandoned by environmentalists, since they drastically altered the site's ecology; and the only way to prevent the precious wetland from drying out was to keep bringing water in from Wicken Lode, after which, in theory, it was held back by the raised banks which encircled the western edge of the site, like dams impounding a raised reservoir of swamp. However, ever since the 1940s, these banks had been badly maintained. Moles, in time-honoured fashion, had riddled them with holes, and the banks were leaking like sieves.

By the early 1980s the result was the worst of all possible worlds. The farmland towards Upware was being flooded regularly by water from the fen, which was speedily drying out. Nobody could agree on how to solve what was evidently a common problem. The bank did not belong to the nature reserve, and the farmer said that it was not his responsibility to repair it either. The Ministry of Agriculture was reluctant to pay for the repair of the bank, because it said that this would benefit the nature reserve. The Nature Conservancy Council was equally reluctant, because it said that it was not its job to improve drainage on adjacent farmland. The opposing camps, described in *Farm in the Fen*, glared at each other across the barbed wire and the thistly bank as the wheat got wetter on one side and the cream comfrey drooped in the desiccating peat on the other. Into this impasse stepped the local land-drainage engineers, whose committee decided to accept responsibility for future repair and maintenance of the bank, specifically because they accepted a remit towards the protection of wetland habitat as part of their duties.

*Historic River Landscapes*

The close involvement of historians from the Museum Service, as well as local ecologists, on some land-drainage projects points to the importance which river managers should be assigning to culture, as well as nature, as they resurrect the landscapes in their care. In our countryside, which is supremely the product of long occupation, the two are generally indivisible.

Thus it is entirely valid for a landscape budget, which should always be included in a river engineering scheme, to embrace the protection and enhancement of the man-made artefacts which contribute to the quality of the river landscape, alongside flowers and birds and trees. Who would have associated the little river Leam in Warwickshire with the opening up of the mighty and mysterious Nile? Yet, on a stone plaque set into the bridge over the Leam at Eathorpe are inscribed the words:

> THIS BRIDGE
> WAS
> ERECTED
> BY
> SAMUEL SHEPHEARD
> OF
> EATHORPE HALL
> AD 1862

Mr Shepheard was able to pay for the construction of this useful memorial to himself out of the proceeds from his famous hotel in Cairo, whence the processions of Victorian explorers set off with their bottles of porter and pale ale into the unimaginable African interior. Having retired from Shepheard's Hotel, its proprietor settled at Eathorpe Hall beside the Leam, which, 120 years later, was subjected to a major land-drainage scheme incorporating all enhancement principles. The plaque on the bridge had become almost unreadable, and was freshly cleaned by the county council, which rebuilt the crumbling bridge in tandem with the scheme. Had they not done so, it is certain that the water authority would have taken on the small cost of cleaning the tablet as a legitimate part of its environmental brief, at the request of the owners of Eathorpe Hall.

On another river in that well-watered county, the land-drainage landscape budget came to the rescue of an altogether less matter-of-fact monument. The weir on the Warwickshire Stour at Honington belongs to the age when such decorous fantasies as Alexander Pope's invocation to the river-gods to rise and acknowledge Queen Anne were expressed not only in poetry, but also in stone:

> The blue transparent Vandalis appears;
> The gulphy Lee his sedgy tresses rears.[13]

Honington weir is adorned by two such crumbling water spirits, whose survival, together with the impounding of the upstream reach of the river through the eighteenth-century landscaped park, was imperilled by the weir's near collapse. Grants for the relatively cheap remedial measures required had been sought in vain, and the water authority, answerable only for specific land-drainage problems, had consistently refused requests for help. With the official designation of a landscape budget for rivers in 1987, however, it was possible to repair the historic weir, and thereby also keep the sedgy banks, which the structure, complete with its reclining deities, retains. In passing, perhaps it should be said that Pope's 'Vandalis' is no longer blue or transparent. It is the wretched Wandle, the foul fluids of whose lowest reaches seep into the Thames at Wandsworth.

## Archaeology

A much less sympathetic approach than that meted out to the Stour, one involving destruction of both our natural and our cultural heritage, could be seen in the summer of 1986 on the Maxey cut between Maxey and Helpston in Northamptonshire: the very waterway whose straight course is the end result of the environmental savaging in the early nineteenth century so lamented in the poetry of John Clare. By the time of the latest operation, it has to be said, the Maxey cut was already an impoverished stream, following centuries of abuse, culminating in massive works which created a new channel in 1953. By now the channel was critically important for flood flows, and the new scheme did include some tree planting. None the less in 1986, from both sides of the bank, yellow diggers reached their long arms down to the water, and, with the precision of gigantic razors, sliced out a cruelly exact trapezoid. The waterside fledge of bur reed and the silvery blowing grasses on the slopes above, which had supported meadow brown butterflies, many moths, and even lizards, were, in the midsummer growing season, removed down to the last green blade. Not even a nettle leaf escaped. The only green was the level wheat, extending from Maxey to Helpston, scored down its middle by the straight, dead cut, to whose sides the machines added a final polish with the flat back of the bucket. In the village, John Clare's monument had been newly cleaned; it might have been a more fitting memorial to him to have taken a little more trouble with the river.

Such scenes are still standard in many parts of the country in the late 1980s. What singles out the work at Maxey as a special kind of ruin was the final destruction of a 5,000-year-old archaeological site, the result of lowering the water-level. The originators of the scheme made little attempt to communicate with—let alone alter their programme to accommodate—the local team of archaeologists, who, in a desperate race for time, struggled to unearth as much evidence as they could, before the inexorable approach of the machines. Wetland archaeology is to dry-land archaeology what ecology is to botany or ornithology. Because water preserves microscopic traces of pollen, plant stems, and the finer scratchings on the surfaces of wood and bone, it makes it possible to reconstruct a much more total picture of the past than that which can be gleaned from

remains preserved under dry conditions. On one wetland site in America, archaeologists were able to identify the blood of different species of animals still
smeared on prehistoric stone tools as being that of bears, sea-lions, and humans.[14]
The desiccation of so much low-lying land in England through drainage is
speedily destroying all this extraordinary evidence. Alongside the Maxey cut,
there was discovered in 1976 a Neolithic camp, 200 metres in diameter, situated
on a long-vanished bend of the river. The archaeologists called it Etton, after
the village nearby. When they began digging in 1981, they discovered among the
pots and antlers, as they peeled back the soft mud, a cache of hazel-nuts, the
spookily preserved mould of poplar leaves which had fallen from the riverside
trees one autumn 5,000 years ago, and the oldest piece of string in England,
made from stems of cannabis or nettle. It soon became clear that the river
engineering scheme in 1953 had triggered a deterioration of the site,[15] which the
pumping operations of the adjacent gravel pit had dramatically accelerated
exactly 30 years later. By 1986, when patient excavation had at last begun to
yield interpretable results, the arrival of the water authority river scheme was
the last straw. By August, when the machines reached Etton, there was no point
digging any further, and the unexcavated portion of the site was abandoned.

The lack of liaison between archaeologists and land-drainage engineers is
certainly a failure of communication on both sides. Archaeologists have still not
organized themselves into the relatively effective lobbies achieved by naturalists,
although the kind of evidence which they unearth has surely as glamorous a
public appeal as any kingfisher or otter. An additional problem is that funding
for excavations is generally available only when sites are already in peril. Attitudes are also changing. In 1987 the Anglian Water Authority came to the rescue
of an archaeological site at Flagg Fen near Peterborough, and the Thames Water
Authority actually employs an archaeologist. None the less, our descendants will
surely blame us for the haphazard way in which river management often treats
the remnants of our ancestors. I will not forget in a hurry the winter day in 1985
when I walked alongside a Shropshire river which had been dredged without
proper thought for the environment, and downstream of which a very ancient
canoe had previously been salvaged, rather by luck than by judgement. As we
stumbled over a giant log crumbling in the frost, the engineer made the immortal
remark: 'Not another bloody Iron Age boat!'

## Creative River Management

Pre-planning, in order to work to the widest environmental brief, is essential for
all land-drainage work, not just those major operations which offer the potential
to destroy or else rebuild the entire surrounding landscapes. Even humble ditches
have their histories and their special sense of place. In the west of England, they
go by the name of 'rhines' (pronounced reen), which in turn led to the local
expression 'rhine wanted', meaning to lie supine. This was first applied to a

Fourteenth-century and twentieth-century engineering. The medieval Old Splott Rhine, bordered with flowering scurvy-grass, flows towards the Severn Bridge.

sheep lodged helpless on its back in a ditch, but soon it also came to describe the legless drunkard, reduced to such extremities by his own excellent local cider, no doubt. Old Splott rhine winds secretly through the Vale of Berkeley in Gloucestershire, under the motorway, and out into the Severn at the point where the new suspension bridge marches across the tidal river with all the might and elegance of which modern engineering is capable. The medieval engineers who built Old Splott had themselves accomplished no mean feat, complicated by the fact that the ditch was the subject of a dispute in 1346 between the lord of the local village of Elburton and the abbot of Malmesbury. The compromise arrived at is reflected in this ditch's present idiosyncratic course. In addition, its gently graded banks change subtly according to the treatment it receives from its various landowners, as well as from the ecological conditions dictated by its varying proximity to the estuary. Some banks are dense with cowslips and primroses, which depend for their survival upon the regular summer mowing carried out

The unacceptable face of river engineering. A watercourse encased in stone and steel. Vale of Berkeley, Gloucestershire.

to ensure Old Splott's primary function as a drain. Its tributary ditches and damper margins are lush with lilac cuckoo flowers; and crowning the banks in places where they do not impede a necessary summer dredge are solid hedges of hawthorn or blackthorn, drenched in April with a snow of blossom, which in autumn ripens to a crop of sloes as dusty with bloom as any grape. Dabchick and moorhen skulk in its little bays of reed, and at its erratic bends, sentinel willows lean drunkenly out over the water. Where Old Splott rhine finally empties its critical flood-waters into the gleaming estuary, the cold salt winds have ensured that its muddy banks are white with the flowers of scurvy-grass, which has even colonized the old brick wall, down which an iron stair leads, for the regular maintenance of a timber flap-valve.

Not all the rhines in the Vale of Berkeley remain like this. A few miles to the north of Old Splott, the ditch systems were massively deepened and widened in the early 1980s, with landscape staff called in *after* the basic design decisions had already been made. The steep sides slipped, and slipped again. Two years and many thousands of pounds later, the ditch system extends to the horizon, a trench of stone and steel. To call the environmentalist in to plant saplings along the crown of this canalized catastrophe is like asking him to ornament a grave with a funeral wreath.

It is important to realize once and for all that such ugliness is unnecessary; that an imaginative environmental approach to the real problems of flood control is very much the art of the possible. Indeed, the new ecological engineering frequently provides the most practical solutions in terms of river management. The past decade has certainly proved to be a turning-point, in that any attempt to reverse the clock and go back to—or, as is still happening in many parts of the country, continue—drain creation, rather than proper river management, will now be made in the teeth of example after example of proven good practice. At so little cost and for such evident reward, margins, pools, and trees can be given back to the brooksides. Flowers can be returned to the flood-banks, and the creative evolution of our river landscapes can be achieved as a by-product of their practical control, at no loss, and, in some cases even an improvement, in drainage efficiency. There are even occasions when a thoughtful environmental approach can actually save money. In 1986–7 the Severn–Trent Water Authority spent a mere 5 per cent of its total annual river-maintenance budget on just such endeavours. If everyone responsible for the maintenance of watercourses rises to such a proportionally modest commitment, then the loveliest of our existing rivers will not be lost, and all the miles of denuded drain which lie waiting, like sleeping beauties, in intensive care will, over the space of a generation, recover life and elegance as they carry away their essential cargo of flood-water to the open sea.

# 8

# The Last-Ditch Stand
## The Wetlands Debate, 1974–1988

THE streams and ditches which feed the river Blackwater in Ireland are sweet and clean. They seep and gurgle through every little valley, among the low hills of Tyrone and Armagh. In winter, the water has always brimmed these valley bottoms: sometimes a silver lake, glimpsed in the distance, like a crescent moon, spread across the wide valley floor; at other times invisible, but soon evident to the walker, wading in the liquid fields, and mobbed by lapwings which swoop in wild delight over their nesting grounds. Some valley corners have been given over to sallow bushes and tussock-sedge. Here, minnow-filled ditches run cold and clear over beds of emerald starwort, delicate as seaweed in a rock pool, and promising, with their mossy winter growth, the washed green brilliance of the coming Irish spring.

What has always held these places in a kind of watery sleep is the level of the river Blackwater. If that remains high, then the soft Irish rain cannot escape fast enough down the myriad streams without spilling over the land. This has led to the presence of more than thirty wetlands adjacent to the middle reaches of the Blackwater. Through long evolution, these have been colonized by curlew, corncrake, cuckoo, and by the cream, waxy flowers of grass of Parnassus and butterfly orchid.[1] In 1986 the mild Irish spring awakened many of these wetlands for what was perhaps to be a last flowering. Dredgers and drag-lines were massively lowering the level of the main river, thereby effectively taking the plug out of the little marshes, bogs, and loughs which lie in the heart of the Blackwater valley and in the valleys of its tributary streams. The vivid evidence of what was about to happen littered the ancient hoar-green landscape. Glossy Kodak-yellow drainage-pipes festooned the countryside, looped and stacked in farmyards, overflowing from the innards of barns, lurking in trailers, and piled in PVC coils in the fields, like gigantic canary-coloured worm-casts. The plastic pipes were ready to begin their work and suck the life out of the wetlands.

Lapwings.

A little wetland of sallows and tussock sedge: a typical target for drainage. Photo: Glyn Satterley.

Drainage pipes stacked and waiting to suck the life out of the wetlands. River Blackwater, Northern Ireland.

Beside the pipes in some places were stored or heeled into the ground generous quantities of young trees and shrubs ready for planting. Old meanders were carefully retained by the design engineers, in whose offices well-thumbed manuals containing all the latest techniques of ecological engineering were in constant use. In the exceptionally trying circumstances of the Irish borderland, an enthusiastic and moving attempt to combine the principles of nature conservation and land drainage was under way. The new orthodoxy was being taken on wholesale—and horribly misinterpreted.

## When Not To Drain

There are two fundamental issues regarding the wise environmental management of rivers. One is creative enhancement, which should always be carried out as part of river engineering, as described in the previous chapter. The other is a fundamental decision which should sometimes be taken not to drain at all, on account of the presence of adjacent wetlands. The wetlands once beside the majority of rivers in the United Kingdom have vanished years ago. While in such

*Top.* A new pond is created. The Hog's Piece, River Leam, Warwickshire. *Bottom.* Wild flowers which are encouraged by the creation of ponds: purple loosestrife (*left*) and lesser spearwort (*right*).

Maintaining the quality of an upland river is not without problems. Fencing is needed against sheep if the trees are to have a long-term future. The downstream reaches of this river system are already treeless. River Tanat, Powys.

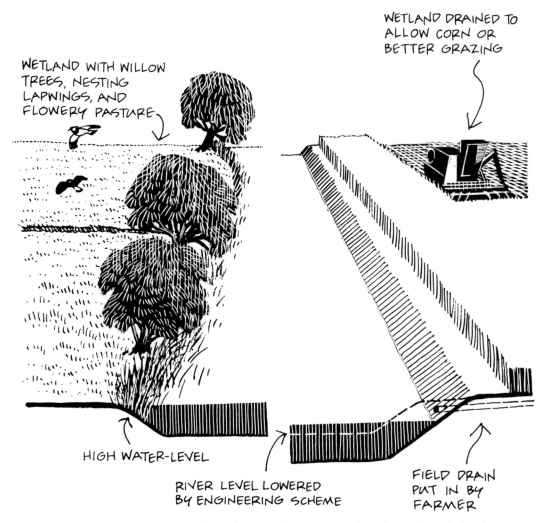

WETLAND WITH WILLOW TREES, NESTING LAPWINGS, AND FLOWERY PASTURE

WETLAND DRAINED TO ALLOW CORN OR BETTER GRAZING

HIGH WATER-LEVEL

RIVER LEVEL LOWERED BY ENGINEERING SCHEME

FIELD DRAIN PUT IN BY FARMER

The principles of land drainage. The left-hand section shows existing damp land. The right-hand section shows the river level drawn down as a result of river engineering, enabling conversion to corn or better grazing.

cases it is possible for engineers and conservationists to strike up a creative partnership, sympathetic riverside management should never become a cosmetic for the drainage of our last remaining wetlands. In those few places where wetlands survive, the conflict of interests is absolute, and inevitable. The purpose of land drainage is to lower the water-level on land adjacent to a river in order to intensify agricultural production. Water lying on the land, which ensures the survival of fritillaries at Cricklade and wildfowl on the Ouse washes, is what the drainers want to expel. 'Waterlogged land' can be a term of abuse on the lips of a progressive farmer, or the ultimate accolade coming from a conservationist. I remember seeing an engineer's slide of a well-known nature reserve, labelled simply 'Bad Drainage'. Somerset farmers used to say of the flowery pastures on West Sedgemoor that the best thing you could do with them was give them to your enemy. In most cases, even with the best will in the world, it is virtually impossible to isolate a wetland from the adjacent main watercourse in such a way as to lower the water-level of the latter, while keeping the former wet. As the flowers, birds, and insects which inhabit these places have been reduced to such a tiny fraction of their former abundance, it is now increasingly accepted that public money should not be spent on destroying wetlands. In addition, many of our best wetlands are now protected by the designation Site of Special Scientific Interest (SSSI) which means that even the site's owner, if he undertakes drainage without consulting the Nature Conservancy Council (NCC), can be penalized. No amount of tree planting, pond creation, or sympathetic channel design can justify a drainage scheme which leads to the loss of fine wetland habitat.

Ten years ago, however, this fundamental principle was far from universally accepted. Direct drainage of wetlands has been the unacceptable face of river management. It has been the nub of the problem throughout a decade of controversy surrounding drainage, and has played a major part in the wider conflict between an expansionist agricultural policy and a national concern to protect the English countryside. The controversy over wetland drainage has sent ministers of state scurrying to some of our remotest and most inhospitable marshes. It has touched on issues far beyond the individual fates of dragonflies and water-lilies, and even beyond the arguments within agriculture. The wetland debate has raised questions about the abuse of technology and the misuse of public money, about the problems of a society brought up against a reduction in the quality of life through its commitment to economic growth, and about the rights of individuals concerning their own property. It has fuelled conflicts between locals and outsiders, between regional and national interests, and has been part of a battle for power between the landed interests of the shires and the finance capitalism of central government.

## By Farmers For Farmers

To understand why these conflicts were inevitable and, indeed, why further conflicts may occur, it is necessary to look at the political and economic position of those responsible for river management within our ten water authorities. Land drainage is the Vatican of the water industry: a state within a state, wrapped in mystique, traditionally the exception when it came to close scrutiny of budgets by authority finance sections, and, most significant of all, answerable only to an autonomous committee dominated by the farmers appointed by the Ministry of Agriculture. This special position is the result of political manœuvring by the farming lobby during the run-up to the 1973 Water Act, whereby the present water authorities were set up. The setting-up of the water authorities as professional multi-purpose bodies was a brave, and in the light of events, generally successful move by the Heath administration. Before 1974 there were some 1,600 authorities responsible for the various aspects of water management. In such

Map showing water authorities and internal drainage boards in England and Wales. Taken from the Ministry of Agriculture map dated 1983.

circumstances, as the pressures of development increased through the 1970s and 1980s, the likelihood of serious water shortages and of pollution problems resulting from antiquated sewage-treatment systems would have been considerable.

Radical reform, with the goal of integrated catchment management, was pushed by Peter Walker in the Cabinet and by the Ministry of Housing and Local Government in Whitehall. However, then, as now, intense rivalry existed between the Ministry of Agriculture, Fisheries and Food (MAFF) and the Ministry of Housing (now the Department of the Environment). Since the days of Vermuyden and before, drainage has always lain at the heart of agriculture, and MAFF officials were vigorous in lobbying to resist what they saw as a take-over by a rival ministry. An alliance of farming interests, known by its opponents in government as the 'MAFFia', suggested that drainage continue to be administered totally outside the new water authorities by the twenty-nine river boards which had previously been responsible for land drainage and which were, needless to say, dominated by agricultural interests.[2] When this proved impossible, they settled for a compromise whereby land drainage retains a uniquely separate organization within the water authorities, serviced by its own statutory committees and with its all-important system of grant aid surviving intact. Jim Prior, the Minister of Agriculture and himself MP for the low-lying constituency of Lowestoft, announced the decision in 1972. One reason for this victory by the 'MAFFia' was its ability to marshal the support of influential back-bench MPs such as Sir Harry Legge-Bourke, at that time MP for the drainage-dominated constituency of Ely and chairman of both the Conservative 1922 Committee and the Association of Drainage Authorities. The threat of delaying tactics which such back-benchers posed to the bill, which was on a tight schedule and in which drainage figured as only a relatively small issue, ensured that the administration of drainage remains to this day substantially unreformed. The story of land drainage has always been that it is low on the agenda. In 1986, when privatization looked like a distinct possibility, river management seemed set to survive yet again as an independent branch of the Ministry of Agriculture in the new order of things, one reason being that to change this long-sanctioned arrangement would have taken a disproportionate amount of very senior ministers' precious time in chairing the relevant committees.

The members of the main board of each water authority are appointed by the Secretary of State for the Environment, with the exception of a maximum of two members appointed by the Ministry of Agriculture. In the hands of this main board rest the ultimate power and responsibility for all the activities of the water authority—that is, with the exception of land drainage, for which the ultimate decisions are taken by each authority's regional land drainage committee. The chairman of every such committee is appointed by the Ministry of Agriculture. Below these committees are a number of local land drainage committees, which cover the different local areas of each water authority and make detailed decisions concerning specific schemes. Both local and regional committees must have a bare majority of appointees from county councils or urban authorities to counterbalance the members appointed by the Ministry of Agriculture, or, in the case of the local committees, farmers appointed by the regional committee.[3] Thus, in theory, the balance of power in these committees is finely poised between farmers and local authority appointees who have less of a vested interest in pushing through agricultural drainage schemes. However, various factors reinforce the farmers' control in many cases. A number of local authority appointees are farmers, and the attendance of those who are not has often been unreliable. In addition, the chairman, always a leading farmer, is in an especially powerful position to steer and lead the committee, and also has a second, casting vote. In December 1985 a parliamentary question was put to the Government concerning the occupations of all those who serve on regional land drainage committees. The written answer is not easy to interpret in all cases, since a number of appointees were registered as either 'retired' or 'occupation unknown'.[4] However, for the committee of the Anglian Water Authority, if the chairman's casting vote is included, farmers held twelve votes, as against non-farmers' five votes, in 1985. In the South-West Water Authority, the farmers could muster eleven votes, as against only three votes from those with other occupations; while in the North-West Water Authority, a slimmer majority of eight votes to six was held by agriculture.

The local land drainage committees are even more universally dominated by agricultural interests, and because they draw their farming members from the areas where drainage schemes are being proposed, they tend to be less impartial in their judgement than the regional committees. Since agriculture and drainage are so closely linked, it is only right and proper that farmers should have a great deal to do with the running of these matters. What is extraordinary, however, is that where a water authority chairman is not in sympathy with large-scale agricultural drainage, he can be excluded from the land drainage committees in his own organization and do little more to influence their deliberations than to heckle from the back of the room. Drainage, financed by all of us, has a profound effect upon our environment; and yet, except for the presence on a few committees of one member with interests in fishing, environmentalists are not represented on the water authority land drainage committees.

Even more startling anomalies in the administration of land drainage are apparent where the direct jurisdiction of the water authorities ends. In England and Wales, all watercourses are designated either 'main' or 'non-main river'. 'Main rivers' are the responsibility of the water authorities, and are in general our most important rivers. All other rivers are lumped together as 'non-main rivers', and can be the responsibility of county, district, or city councils, or can be engineered by organizations and individuals ranging from the National Coal Board or the Ministry of Transport to private owners. While it is unreasonable to expect anyone to define exactly the point at which a river or its tributary changes from harmless brook to potential flood problem, the map of 'main rivers' in England and Wales looks remarkably inconsistent in some areas.[5] In the agricultural lowlands of Anglesey, Essex, and the Lancashire Fylde, almost every ditch and drain appears to be marked as a 'main river', while the river Rea, 'the mother of Birmingham' and principal watercourse of our second city, remains unmained.

### Internal Drainage Boards

In the lowest-lying land of all, and, therefore, exactly in those places where our surviving wetlands present temptations to traditional drainage men, drainage is administered by some of the most archaic public institutions to survive substantially unreformed into the late 1980s in the United Kingdom: the internal drainage boards (IDBs). These boards, whose origins date back to medieval times in some cases, were all brought under a single piece of legislation by the 1930 Land Drainage Act, which preserved much of their autonomy intact. The idiosyncratic constitution of the Adlingfleet IDB dates back to the late eighteenth century, while the board which manages the internationally important wetland on South Lake moor in Somerset is run according to rules laid down in 1830. Internal drainage boards are single-purpose bodies set up with the principal aim of draining wetland for agriculture and maintaining drainage on previously drained lowland. Only occasionally, as in the case of Spalding, does an IDB look after urban drainage. In 1982 there were 248 such boards, which, by 1986, through amalgamation, had shrunk to around 225.[6] They occur in such places as the Fens, the Broads, the Vale of York, the Somerset Levels, and Romney Marsh. In such lowlands, the major watercourses are managed by the water authorities, while the principal drains which feed into those watercourses, together with the pumps, are the responsibility of the IDBs. The smallest ditches of all are maintained by the farmers on whose land they occur. In a classic assault on a wetland, the water authority will turn the key by lowering the level of the main river, and the IDB will then move in to co-ordinate and control the pick-up drainage. The farmers who sit on the local land drainage committees of the water authorities are frequently the chairmen of the internal drainage boards in the same area. The members of the boards are almost always farmers, and, more than that, generally the largest and most progressive in the neighbourhood. In a House of Lords debate in 1981, Lord Buxton made the point:

In the good old days there was no problem with drainage boards. It was then sensible to have the local farmers and people who owned the land on the local land drainage boards. It made reasonable sense that the clerks of the boards were often their solicitors. But times have changed. . . . Machines can do work in a morning which used to take 20 men a week or a month.[7]

The environmental consequences, when such organizations can employ drainage contractors with heavy plant, or, in the case of the larger boards, when they possess their own fleet of machines and can buy large new pumps, are profound. Internal drainage boards are the parish councils of wetland management, organized exclusively to serve parochial interests; but, unlike parish councils, they are authorized to levy rates and have access to massive grant aid from the Ministry of Agriculture, and, most significant of all, their activities can and do affect a national, and even an international, resource, the flowery pastures and winter washlands of our major remaining wetlands.

A common defence of IDBs is that their small, locally based character makes them relatively self-supporting and cheap to run. However, they are eligible for Ministry of Agriculture grants of up to 26 per cent of capital costs, and have received between them, on average, £2.5 million a year in this way. Through their special powers of levying rates, they also raise approximately £10 million annually.[8] The rating rules by which they secure their money are remarkable. First, their jurisdiction coincides with that notoriously erratic, unprovable, and non-statutory cut-off point known as the Medway Letter Line, which was defined in 1933 as the point which lies 8 feet above 'known' flood-level.[9] In theory, the only ratepayers living below that line are the owners of flood-prone agricultural land. In reality, however, power-stations all along the flood plains of the Trent, the Don, the Aire, the Yorkshire Ouse, the Soar, and the Medway are far and away the largest contributors to the finances of their local IDBs. In one case, a power-station is recorded as having contributed over 90 per cent of a board's income.[10]

Typically, Laneham IDB on the Trent has been able to carry out all its major works on the back of contributions from the Central Electricity Generating Board. In north Kent, the Sheerness dockyards and the oil refineries along the Thames estuary boost the local IDB's income. The south Gloucestershire board is fortunate in having within its region a veritable gold mine in the Avonmouth industrial estate, much of which is owned by the Bristol City Corporation. Thus, indirectly, the citizens of Bristol are contributing to the deep drainage of part of the Vale of Berkeley. Power-stations and industrial estates below the Medway Letter Line gain very little return in flood protection from their IDB rates. The construction and maintenance of sea-walls and the control of major outfalls to keep them flood-free is the responsibility of the water authorities, to whom they pay a separate set of drainage rates. What they are financing through their IDB rates is primarily agricultural field drainage. The same is true of those low-lying towns and villages which pay an IDB drainage rate. These include parts of York, Hull, Goole, and Boston, as well as some villages which depend upon tourism, and for whom, therefore, neighbouring wetlands, which draw bird-watchers and holiday-makers, may be an important economic resource. Thus Burnham-on-Sea contributes to the lower Brue IDB, which played a major part in controversial attempts to drain the Somerset Levels; while Caister-on-Sea, a tourism-dependent village near the Norfolk Broads, contributed 50 per cent of the cost of a scheme carried out by the Muckfleet and South Flegg IDB which canalized the lower reaches of the river Muckfleet in the early 1980s for the benefit of agriculture.

Furthermore, agricultural land has never been re-rated since 1935, whereas urban property has been re-rated a great many times since then. Consequently, farmers, for whose benefit IDB operations are intended, pay proportionally far less than urban ratepayers. Lord Buxton explained to the House of Lords in 1981 that in one IDB area, 'the average rate per acre for agricultural land is 23 pence, and for non-agricultural property (wait for it, my Lords) is £165.91p.'[11] The boards are empowered to vary the rates each year, and can levy a lower or zero rate on any owner or occupier 'for any reason'.[12] This does not mean that town-dwellers and small farmers receive a lighter burden, however. If anything, the reverse is true. Little or no effort is made to ensure that the most productive and well-drained land pays accordingly higher rates. When drainage improvements are made, the board can decide both the level of expenditure and the period of time over which that money must be spent. Thus a sudden increase in drainage rates to pay for a major new scheme which converts grazing marsh into arable land may be easily affordable by a large farmer, who has access to both capital and such expensive equipment as combine-harvesters to recoup the benefit, while driving his small neighbour who works on a tight margin as a grazier out of business. Thus it is that the power to manipulate water-levels held by the members of an IDB enables them to control not only the fate of the wildlife in the wetlands, but also the livelihood of the local farming community as well.

And who are these men? Now we come to the most Dickensian part of the entire story. In contrast to other forms of English public life, where such practices were abolished by Clement Attlee in the 1940s, voting strength in the IDB election is proportional to the amount of property which an individual owns. Local residents are permitted to vote only if they own property with a rateable value exceeding £30. Above this, the number of votes which a landowner can cast is determined by the value of his holding, up to a maximum of twenty votes for owner-occupiers, although, in practice, some limits are placed on the number of votes available to one individual.[13] In practice, however, generations can pass before some IDBs ever hold an election. Members are simply co-opted on to the executive, and in this way the normally convenient rule which encourages the largest landowners to dominate the board can be broken when it does not suit the prevailing agricultural interest. The Royal Society for the Protection of Birds, although now the major landowner on West Sedgemoor in Somerset, and owning land on the Norfolk marshes, has not been allowed representation on the appropriate IDBs. Lord Buxton has described these boards as 'self-perpetuating clubs', and even a report commissioned by the Ministry of Agriculture which was favourable to them admits that 'there did seem to be a tendency too for existing members to bring in their friends and acquaintances'.[14] A typical arrangement on some boards is for a large farmer to be present, together with his farm manager and another farm employee. Along with this balance of power which favours the largest landowners come some very special privileges indeed. Boards have the power to enter any land within their boundaries to dig away at meadows and dykes on seven days' notice, and regardless of the owner's wishes. In this way, they have entered two reserves in Norfolk, together with Martham Broad, all with the express aim of draining parts of these sites.[15] In 1988 an IDB damaged an SSSI at Lakenheath without seeking any permission. It is not uncommon for the engineer or senior partner of the firm of consulting engineers which supervises drainage for an IDB to hold the office of clerk and sometimes even treasurer for that IDB. The opportunities which this presents for vested interest to promote large-scale drainage are easy to imagine.

None of these arrangements would matter so much if we were all still digging for victory. Indeed, if there were an overriding national need for agricultural intensification, IDBs would present a worthy, efficient vehicle for wetland management. The problems have arisen because the whole thrust of these organizations has often been geared towards intensive drainage, which has ceased to be nationally essential. This has led to conflicts between large and small farmers, as much as between farming and conservation. The relationship between IDBs and conservationists has developed from one of mutual incomprehension to one of mutual hostility, with leading figures in IDB organizations likening conservationists to 'Argentinians'[16] and a popular paperback on conservation labelling the IDBs as 'infamous'.[17] More recently, a welcome initiative by the Association of Drainage Authorities has begun to foster closer co-operation between conservationists and IDBs, and in a few cases, notably the Ouse washes, conservationists and farmers work closely together on the same board (although the nature of the Ouse washes as essential washland ensures that it could never have been subject to serious drainage pressure anyway). Engineers on some of the Fenland IDBs manage their ditches in an environmentally sympathetic way, but here, too, any significant wetlands have either been drained or are fully protected. Finally, younger farmers, whose formative years were not dominated by the farming slump of the 1930s and the food crisis of the War years, are just beginning to gain chairmanship of the IDBs.

## Attitudes to Drainage in the Mid-1970s

This was less true of the IDBs and the water authorities in the mid-1970s, however. When the water authorities were set up, with unprecedented resources in money and staff to initiate what seemed to some the golden age of drainage activity, the senior men, then at the peaks of their careers in the water industry and the Ministry of Agriculture, could remember only too clearly the times when Christopher Addison, Ramsay MacDonald's Minister for Agriculture, had described the rural dilapidation in which 'the fields often enough, with their vistas of weeds and rubbish, cry aloud for land drainage'.[18] Thus a senior engineer with responsibilities for drainage in the Ministry of Agriculture stated in 1979 that 'Agriculture, Britain's most competitive industry within the European Community, is assailed from two sides. Vast quantities of good agricultural land are lost each year to urbanisation, whilst on the other side, the sea eats away more of this industry's raw material.'[19]

Agricultural economists have shown that, in reality, the loss of farmland to urban sprawl peaked in the 1930s; and not only does most of our agricultural land in the shire counties remain intact, but in general, past losses have been compensated for by vast improvements in agricultural productivity.[20] In 1975, the year after the water authorities were formed, the Government brought out its influential White Paper 'Food From Our Own Resources',[21] which encouraged all forms of food production as an entirely public-spirited endeavour, especially in order to reduce our balance of payments, a problem which is now considerably diminished due to the advent of North Sea oil. In those heady days, it was not uncommon to hear land-drainage engineers talk of the need to build up supplies of food in case we should ever again suffer from a blockade—a fantastical notion in the age of the nuclear bomb. Many water authority members of staff were taken on from the old river boards, and it was to be a long task for more enlightened senior management to re-educate some of them to understand that they were now the servants of the public at large, rather than simply of agriculture. Meanwhile, the price of land continued to soar, as witnessed by the increasing investment in farmland by financial institutions; and this provided a major incentive for farmers, both at the grass-roots level and in the upper echelons of the water authority committees, to call for further drainage. Furthermore, serious flooding was fresh in most people's memories. In July 1968 heavy rainfall coincided with a high spring tide on the Bristol Avon. Overnight, more than 3,000 houses and shops in south Bristol were inundated, and when the flood-water retreated, it left in its wake a noxious layer of sewage and stinking silt. Doctors, studying the health of the flood victims in the aftermath of the disaster, concluded that there was a 50 per cent increase in deaths in the year following the flood, together with increased surgery attendances and psychiatric illness.[22] The legacy of the flood, in terms of the long-term effects of stress, was added to the immediate destruction wrought upon life and property. By the late 1970s, with an army of engineers and machinery controlled by a drainage lobby which could also muster immense political and administrative power, everything seemed set to tame the flood once and for all.

On the other side of the drainage ditch, the defenders of wetland habitat and those who questioned the virtues of wholesale drainage were in serious disarray. Most fundamentally, a total lack of planning control exists on even the largest land-drainage scheme, as on most other agricultural activities in the countryside. Before 1981, effective protection for the most precious areas, Sites of Special Scientific Interest, did not exist. The 1973 Water Act suggested that water authorities should merely 'have regard to the desirability of preserving natural beauty, flora and fauna . . .' which must stand on record as one of the most half-

Butterbur, commonly found along riversides especially near water-mills.

hearted injunctions ever devised by an ingenious lawyer. Within the water authorities, while staff and money were available to promote recreational amenities, it did not occur to anyone at first to employ people with a specific remit to safeguard the habitats which are so profoundly affected by the activities of the water industry. Even in 1986, only half our water authorities employed trained staff with particular duties regarding conservation.[23]

Outside the water authorities, the organizations responsible for nature conservation were woefully undermanned. The NCC, that branch of government charged with owning and leasing national nature reserves, as well as designating SSSIs, was so understaffed that, in 1983, it enjoyed the unique experience of having its staff *increased* by the stringent Rayner Review, through which the Government investigates and then normally axes public-sector departments. A small, but indicative, example of the permanent resource crisis at the NCC was the way in which at one stage in the early 1980s, those seeking advice from its scientists often had to collect and deliver them to their destinations, since they were allowed neither cars nor mileage allowances. Under these circumstances, water authorities requiring survey information on wetland habitat which might be threatened by proposed drainage schemes often had to turn to county trusts of planks, pieces of old door, and a daub of mud—all that holds in the wetland wildlife of the northern washlands of the Poitevin. This sorry scene was visited by Madame Bouchardeau, then Minister for the Environment, and the powers in Paris are now debating the future of the Marais Poitevin. Its fate is more likely to depend upon the legal technicalities of farmers' rights to drain and upon the outcome of a political battle between the Ministry of the Environment and the Ministry of Agriculture than upon any real merits of the case, however.

### Amberley Wildbrooks

One major difference between the French and English environmental lobbies is that English conservationists, although ill organized initially, have had at their disposal a national culture which is steeped in landscape, a concept and a value which was virtually invented by the English and bequeathed by us to the rest of the world. Reared on the tales of Beatrix Potter, *The Wind in the Willows*, and *Watership Down*, the average Englishman will respond to the misfortunes of animals and birds with passions which are sometimes slower to surface when it comes to the tribulations of his fellow men. This is especially true in the south of England, where an articulate middle class has arrived in the pretty villages of Kent and Sussex specifically to enjoy the delights of the countryside, and frequently without any economic dependence on the farming or local industry which might damage that countryside. It was here, at Amberley in West Sussex, that the first serious engagement between a water authority and conservationists on a wetland issue took place, in the new circumstances of agricultural intensification resulting from our membership in the Common Market.

The village of Amberley has everything which a seeker of rural retirement in Sussex could hope for. Thatched cottages wind along a ridge towards a Norman church and a fourteenth-century castle, a quiet echo of the fantastic pinnacles and battlements of Arundel Castle further down the valley. Amberley Castle commands a bluff above the 900-acre basin of the river Arun, which, cradled in the lap of the South Downs, is known as the Amberley Wildbrooks. Here, a maze of dykes, which support more than half the species of our entire aquatic flora and seventeen species of dragonfly, converge on the serpentine loops of the Arun. One of the great specialities of this place was always the large flocks of Bewick's swans, which arrived from Siberia to winter on the flooded meadows. Viewed from the hill, they resembled distant herds of sleek white geese, gathered in the inaccessible heart of the Wildbrooks and bringing with them an improbable flavour of the Russian steppe to the domesticated landscape of the Home Counties.

Following the wet winter of 1974–5 and a subsequent chorus of complaints from farmers, especially those wishing to intensify grazing on the flood plain in order to release the upland areas of their land for arable cultivation, the Southern Water Authority applied to the Ministry of Agriculture for £245,000 towards a

Some rivers and wetlands subject to drainage schemes in the 1970s and 1980s.

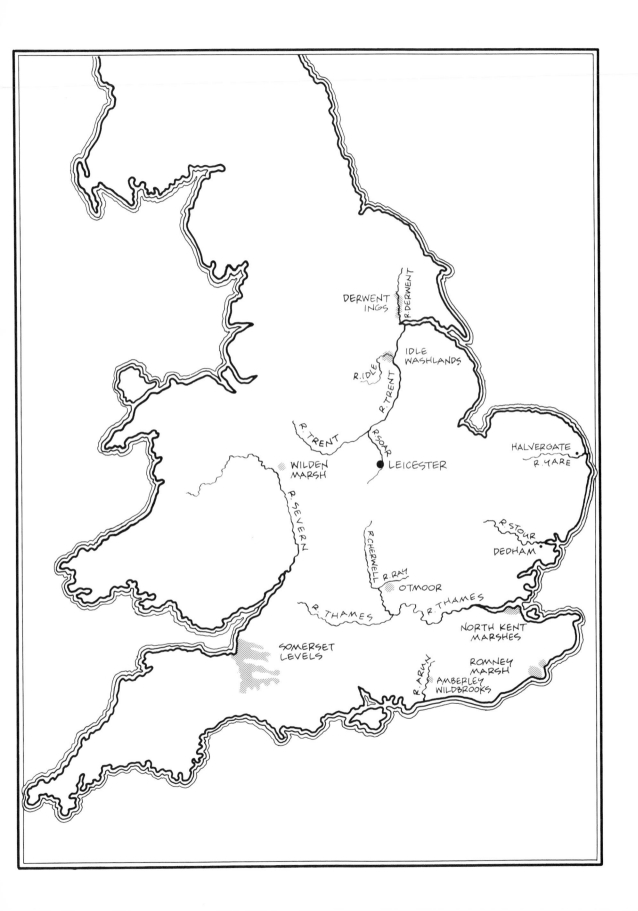

DERWENT INGS

R.DERWENT

IDLE WASHLANDS

R.IDLE

R.TRENT

R.TRENT

R.SOAR

WILDEN MARSH

HALVERGATE

R.YARE

●LEICESTER

R.SEVERN

R.STOUR

R.CHERWELL

R.RAY

DEDHAM

OTMOOR

R.THAMES

R.THAMES

NORTH KENT MARSHES

SOMERSET LEVELS

ROMNEY MARSH

R.ARUN

AMBERLEY WILDBROOKS

Bewick's swans.

£340,000 scheme designed to pump the Wildbrooks dry.[33] The local amenity lobbies mounted an energetic campaign of protest, backed by national bodies such as the Council for the Preservation of Rural England and, above all, the Nature Conservancy Council. In 1978, John Silkin, Minister of Agriculture, took the unprecedented step of setting up a public inquiry on the proposed drainage scheme. In this inquiry, two important points were made. First, natural communities were acknowledged as being irreplaceable; and, second, the economic case put forward by the improvers was said to ignore the fact that the total community, namely the residents of Amberley, who helped pay for the drainage, was suffering an overall loss in its quality of life through the destruction of the Wildbrooks. The NCC's chief advisory officer explained: 'This is a test case. . . . If it became known that we sacrificed this outstanding wetland for the sake of a minute increase in national food production, Britain would lose credibility among the nations who rightly expect us to take a lead in the matter of rational use of natural resources.'

Leeds economist John Bowers, opposing the drainage, made the critical distinction: 'Where society foots the bill, it is the returns to society that matter. It is therefore vital to distinguish social costs and benefits from the private costs and benefits of individuals.' His close examination of the Southern Water Authority's cost–benefit analysis revealed an overstatement of the benefits in economic terms, even leaving aside the environmental issue. The inspector upheld this view, and the pump-draining of the Amberley Wildbrooks was prevented.

In allowing environmentalists detailed access to the figures in this case, the cat had been let out of the bag. It was to be almost a decade before the water authorities and the Ministry of Agriculture relaxed their rules of secrecy concerning cost–benefit calculations for land drainage. Seven years later, the farmer and Tory MP Kenneth Carlisle commented on the confidentiality shrouding costings of large drainage schemes:

The issue of secrecy is especially relevant for drainage. . . . In fact the relevant information often was available before the Inquiry in March 1978 . . . at Amberley Wildbrooks in

Sussex. This found that the cost–benefit analysis was too optimistic. . . . The scheme was thus rejected. Since then it seems that information is guarded more closely.[34]

Furthermore, despite strenuous requests by conservationists for full public debates on other wetland issues, agriculture has never again allowed itself to be dragged into a public inquiry on a river improvement scheme. Amberley set an important precedent on the issue of when not to drain, but in many ways it was a false spring for conservationists. What is more, for bird-lovers it represents the victory that never was, for although abandonment of the pump-drainage scheme saved the flowery ditches and the pastures where lapwing breed, the gravity-drainage which was agreed upon has greatly diminished winter flooding, and so large flocks of wild swans no longer fly in to roost in the lee of the Sussex Downs. The Wildbrooks was also an exception in being situated in a part of England where amenity lobbies are especially powerful. Conservationists were able to orchestrate protests from as far away as Hong Kong, New Zealand, and Germany, as well as from all over the United Kingdom. The *Farmers' Weekly* reported the views of one farmer on the Wildbrooks, resigning himself to another short grazing season: ' "All the nature conservancy bodies are to be stopped from coming on our land." He claimed that major conservation groups had singled out the Amberley scheme as a test case. "There weren't many of us to oppose them and we were not wealthy or well fixed." '[35]

## North Kent Marshes

This may have been the case with some of the farmers at Amberley, but it could scarcely be said to be true of the agricultural lobby as a whole. This was now able to turn its attention to other wetlands, which lay less in the public eye and further from the doorsteps of those who knew how to muster effective opposition in their defence. Just as London turns its back on many reaches of the Thames, so the wealth and influence of Kent has tended to be concentrated in the lush Weald and the warm south, facing away from the county's most spectacular topography: the great estuaries of the Thames, the Swale, and the Medway. The north Kent marshes have had their admirers, however, principal among them Charles Dickens. Dickens spent six impressionable years of his childhood at Chatham, and in later life he returned to that part of Kent, known as the Hundred of Hoo, making the desolate marshes at Cooling the setting for the memorable opening of *Great Expectations*. Here, the convict Magwitch, escaping from the prison hulk moored in the estuary, emerges on a bleak winter evening to terrify the infant Pip, as he becomes aware that 'the dark flat wilderness beyond the churchyard, intersected with dikes and mounds and gates, with scattered cattle feeding on it, was the marshes; and that the low leaden line beyond was the river; and that the distant savage lair from which the wind was rushing was the sea'.[36]

These marshes still retain some of the quality of wilderness which so often

The north Kent marshes.

lured Dickens and his friends, walking over the stubble fields from Gadshill, to revel in the atmosphere of river mist and solitude. Out in the marsh, you can still see the long low batteries built to defend the Thames against Napoleon and the ruined cottages of the decoy men who, in Dickens's time, fortified by Dutch gin and laudanum, made a hard, but independent, living by trapping wild duck. Twenty years after Dickens's death, Denham Jordan, writing under the pen-name of 'A Son of the Marshes', described their life-style and their hunting-grounds as those

most treacherous parts of the flats, covered with splashes and pools, with a film on the surface of them, coloured with all the hues of the rainbow. . . . Here the Bittern hides in the daytime, stirring himself when the Marsh Owls are on the hunt, when thousands on thousands of frogs make the whole place appear to tremble with their croakings.[37]

In Jordan's time the elegant avocet bred in north Kent, and the estuary was a paradise for birds, whose wild calls were believed by marshmen to embody the souls of those who had been drowned at sea. Even before the end of the century, however, Jordan records how an 'old shooter, a man who until then, had lived by his fishing and his shooting . . . said his living was gone, so it did not matter how soon they put him underground. The folks were all mourning the draining of the marshes.'

The draining, which was greatly accelerated in the 1970s, has eliminated many breeding birds from the marshes, as well as one of the estuary's special characteristics: a curious quality of belonging neither to the land nor wholly to the water. There remains, especially at the mouth of the Medway, the spectacle of an immense estuary with its archipelagos of islets to which wildfowlers still cautiously make their way at low tide, Deadman's Island, Slaughterhouse Point, and Bedlam's Bottom. At Egypt Bay, where Dickens loved to walk, the prison-ships are gone, but tankers from the London Docks, their high sides towering above the levels like floating power-stations, glide by on the main stream, while the flares from the oil refineries wink from the distant Essex shore. At Chetney, the tarred ribs of abandoned barges are sunk in ooze, and pear-trees, weighted with woody olive fruit, their bark scaled and crazed like a crocodile's back, descend the hillsides to the edge of the marsh. What is going or gone from the magnificent framework of this landscape is the fine detail which perfected it. The soupy meadows where the gurgling redshank browsed, which extended unbroken between the orchards and the sea, have given way to spanking new ploughed fields of winter wheat.

What that ploughland has replaced was not dead-level grassland, but uneven grazing marsh, which had gradually been reclaimed since Roman times, in which open pools, damp hollows, and patches of reed provided an ideal variety of habitat. The marshes were pitted with borrow dykes, dug out for collecting spoil to patch sea-walls, and intersected with winding ditches known as fleets, which ranged in size from wide lagoons to narrow creeks, colonized by specialized crowfoots adapted to the brackish conditions. In addition to the complex nature of the pasture, where waders nested and wigeon grazed, the total jigsaw of the marshland habitat included mud-flats, which serve as rich feeding grounds for waders, and open water, ideal for roosting duck.[38] For this reason, the Royal Society for the Protection of Birds put forward the north Kent marshes as early as 1942 as one of our most important havens for water birds nationally. In 1963 these marshes were described as 'the most outstanding wildfowl haunt in Kent'.[39] Even in 1982, the surviving area of the north Kent grazing marsh supported a quarter of the total breeding redshank surveyed in England and Wales.[40] This wealth of surviving birdlife was all the more import-ant because so much of the Essex marshes across the water and Romney Marsh in the south of the county had been ploughed during or just after the War.

Redshank, a major casualty of drainage in the north Kent marshes.

effectively halted the ploughing of surviving marsh within the special sites. There has been controversy in the Press about the high levels of these payments in north Kent,[42] especially in cases in which drainage on the Thames estuary is manifestly failing in straightforward farming terms, as deflocculated clay clogs the pipe-slots and the cracks and fissures in the soil.[43] These high compensation payments certainly reflect the remaining major problem of high incentives for cereal production available under the Common Agricultural Policy. However, with no further ploughing since 1981, grassland agriculture, rather than an overspill of industrial London, offers the best hope for the surviving birdlife of the Thames and Medway estuaries.

## Wetland Destruction in the Midlands

In 1980–1 the Severn–Trent Water Authority completed a major scheme on the river Idle in north Nottinghamshire. This is the southern end of the great fen tackled by Cornelius Vermuyden in the early seventeenth century, known at its northern reaches as the level of Hatfield Chase. The Idle and Misson washlands, which were the main target of the scheme, were among the last areas of grazing in a predominantly arable area, and were largely owned by pension and insurance funds. Every winter when the river Idle overtopped its banks, the flooded grassland attracted large numbers of Bewick's and whooper swans, averaging 80 Bewick's in most winters, and in 1971 totalling as many as 174 birds, giving the washlands a status of international importance. When the NCC objected to the drainage scheme, the water authority conceded some small compromises, dropping certain areas from the scheme and offering to create a shallow pond known as a 'scrape'. Although the main washlands affected had already been designated an SSSI, the NCC capitulated immediately, and there was little or nothing which voluntary organizations such as the Royal Society for the Protection of Birds could do to counteract the decision. The SSSI on one of the lowest washlands, Misson East, was accordingly reduced from 168 acres to 12 acres, and the other little areas retained were too small for birds to settle securely without disturbance from walkers. The number of swans dropped to seventeen in 1981–2, and finally to none at all the following winter. The Lea marshes in Lincolnshire, which took a few of these displaced swans, now also suffer disturbance pressures from easy access resulting from drainage works carried out there. In 1986, the NCC commented: 'The prospect is that many birds will be unable to sustain themselves during harsh periods in the Lower Trent Valley and their populations will drop.'[44]

If the Idle scheme had been suggested five years later, there would have been little chance of its getting approval in the changed atmosphere of the mid-1980s; but reversing the situation now is another matter. Engineers are considering raising water-levels a little, but taking farmland from private landowners who are now used to a profitable annual crop is easier said than done. At the same time that the Idle scheme was going through, the Pevensey Levels in Sussex were threatened by pumping proposals, and 3,000 acres of flood meadow on the Nene washes near Peterborough were imperilled by a scheme proposed by the Anglian Water Authority. In 1979 the 144-acre Wilden Marsh, a fine area for marsh orchids and marsh cinquefoil and the most important site for breeding snipe in Worcestershire, was drained.

## Wildlife and Countryside Act, 1981

With such an accelerating attack on our remaining wetlands, it was not difficult to see that a serious crisis was besetting the countryside. Drainage was only part of a wider problem of destruction, in which, in 1980, the NCC estimated that almost 12 per cent of our SSSIs were vanishing every year.[45] These included areas of deciduous woodland and upland moors which were being eliminated by the activities of farming and forestry. If these sites, the cathedrals of the natural system, were disappearing so quickly, the chances of our being able to hand over a landscape of any quality to our children seemed increasingly remote. Concern over the inadequate protection of these sites and, in particular, over the ploughing of large areas of Exmoor led the Labour government to introduce a Countryside Bill in Parliament in 1978. This did not survive the crumbling Labour administration, but in 1979, upon taking office, the new Tory government announced its intention to introduce its own Wildlife and Countryside Bill.

For an issue which traditionally had been above party politics, the cause of nature conservation provoked long and bitter debate in both Houses.[46] In October 1981 the Wildlife and Countryside Act received the royal assent, following several hundred hours of parliamentary time stretching over 11 months, during which a staggering 2,300 amendments were tabled. The central issue was *how* to effectively halt the destruction of SSSIs—through enforceable planning controls or by voluntary means and goodwill, as advocated by the National Farmers' Union and the Government. The Labour opposition, eagerly identifying with popular concern for imperilled wildlife, threatened to talk the bill out unless the NCC was empowered to demand 3 months legal notice from all owners of special sites intending to carry out potentially damaging operations on them. In this way, for the first time, conservationists obtained a real say in agricultural land-use planning. The pressure of time which helped the conservationists win this compromise from the Government was made more urgent by the impossibility of extending the parliamentary session, due to the forthcoming royal wedding. Thus do the accidents of history have unexpected consequences, even for the survival of badgers and butterfly orchids.

The compromises made were by no means all in favour of conservation, however. Whenever the NCC notified an SSSI for the first time, it was now required to give the landowners three months advance notice. This notorious arrangement, known as the 'three-months loophole', gave landowners the opportunity to destroy a site with impunity before being subjected to any restrictions. It acted as a positive incentive for owners to destroy potential sites, and as a major disincentive to the council to declare sites of interest, for fear of their immediate destruction. The most controversial measure of all, however, was the provision whereby the council was expected to compensate landowners whose applications for an agricultural grant were turned down on conservation grounds. This was rather like refusing a city developer permission to knock down an ancient church to make way for a multi-storey car-park, and then paying him the profits which he would have made from that car-park out of the public purse. The most worrying thing of all was the particular public fund which was to be used for this compensation. The director of the Council for the Protection of Rural England, writing to *The Times*, explained that the new Act

gives legal expression to the surprising notion that a farmer has a right to grant aid from the tax-payer: if he is denied it in the wider public interest, he *must* be compensated for the resulting, entirely hypothetical 'losses'. . . . There is, however, even greater cause for concern. The Bill requires compensation to farmers to be paid not by the Ministry of Agriculture, whose relentless promotion of new farming methods through the grant system is now the source of many conflicts, but from the meagre budgets of conservation agencies such as the NCC and the National Park Authorities.[47]

One last good thing came out of the 1981 Act, however. This is clause 48, known as 'the enhancement clause', which has been especially useful in providing a legal charter for the kind of sympathetic river engineering described in chapters 6 and 7. Section 48 requires that water authorities and internal drainage boards 'shall . . . so exercise their functions . . . as to *further the conservation and enhancement* of natural beauty and the conservation of flora, fauna and geological and physiographical features of special interest' (my emphasis). Objections were raised against this clause at the time, similar to those which successfully scouted an attempt to put a parallel 'duty to further wildlife conservation' on the Ministry of Agriculture as part of the 1985 Wildlife and Countryside Amendment Act. In the experience of managers in the water industry since 1981, the worries under-lying these objections have proved groundless. The first objection was that, since good practice was being carried out by many river managers anyway, there was no need to formalize this arrangement. The short answer to this was that if good practice was so universal, then the water industry had nothing to fear from the clause. The second objection to the enhancement clause was that it would be unenforceable. Of course, as with so much legislation, the whole point is that it cannot be policed in all cases. But the beauty of the enhancement clause is that, especially as far as public servants are concerned, it acts psychologically as a charter for river engineers, giving them a legitimate brief to protect, and even go one stage further and creatively evolve, the rivers in their care. Naturally, there are those who interpret the injunction 'to further the conservation and enhancement' of habitat in a far more progressive way than others. What the

Act provides is official justification for the high standards which some of them are setting for their colleagues, so that we can look forward to a steady process of civilizing the rivers, whereby best standards increasingly become accepted as normal standards. One way to change the world is by setting precedents, which demand, in turn, an ever higher standard of public expectation. The open-ended wording of the enhancement clause, while allowing less enthusiastic engineers to get away with a bare-minimum approach to their environmental responsibilities, also leaves room for more progressive river managers to raise the standards of good practice.

On the vexed issue of when not to drain, however, water authorities and internal drainage boards were far more reluctant to embrace a rounded environmental brief. Many of them pointed out that the enhancement clause applied only in so far as it was consistent with their other duties; and in 1981 there was no lack of engineers and farmers who still saw themselves as invested with a divine right to drain. The battle for the wetlands, as opposed to the case for better river management, was to turn on the availability of sufficient funds to compensate those who forewent the benefits of drainage, together with the resolve with which the NCC used the limited powers which, after such laborious effort, had finally been delivered into its hands. On these counts, the future of wetlands in the early 1980s looked bleak.

The government estimate of the future cost of management agreements, by which the NCC was to pay landowners not to destroy SSSIs, at the time the bill was going through parliament was '£700,000 on average per year',[48] a hopelessly unrealistic figure in the light of calculations made in 1984, which estimated that the annual bill for management agreements could be up to £40 million, two and a half times the council's total grant for that year alone.[49] The NCC also faced the mammoth task of renotifying some 4,000 SSSIs in accordance with the new Act, an overwhelming consumer of all its slender staff resources, especially in the years immediately following the passing of the Act. But the Government, the National Farmers' Union, and to some extent the NCC itself had nailed their colours to the mast of voluntary co-operation. It remained to be seen whether either side could contain its more belligerent members. The National Farmers' Union exhorted farmers to be on their best behaviour, although there was a feeling even among its leaders, that, if driven too far, the agricultural community would 'rip up . . . hedgerows out of sheer frustration and annoyance'.[50] The NCC had, at its back, increasing pressure from the voluntary wing of the conservation movement, which, during the passage of the Act, had learned a great deal about the politics of direct confrontation, the lobbying of M Ps, and the value of appeals to the media. The Wildlife and Countryside Act was on trial. Both sides knew that the way the legislation was interpreted in practice in the first few years would dictate the level of restrictions put on how farmers could earn their livelihood, and would also decide whether there really was a place for nature in the countryside of the future.

*Romney Marsh*

The first move in the changed circumstances of the wetland debate took place where some of the earliest traditions of land drainage had been recorded in the Middle Ages: Romney Marsh. Known locally as 'the Marsh', its keen winds and wide spaces are a perfect antidote to the comfortable countryside of the adjacent Kent and Sussex Weald. The solitude and eccentricity of this place, with its hamlets still much as Cobbett described them—'a village with five houses and a church capable of containing two thousand people!'[51]—have captured the imagination of modern painters and photographers, including Paul Nash, John Piper, and Fay Godwin. John Piper, with his accurate artist's eye, has best described the visual quality of Romney Marsh:

In winter the reeds still blow in the dykes showing a pale-yellow flank of close stalks. . . . The fences . . . grow an emerald lichen that at first seems to dust and finally screens the pale grey wood. . . . They look excellent against the blackthorn hedges in winter, when these are dark purple-brown, and against the oranges and reds of massed willow branches. The levels themselves are often paler than the sky.[52]

Romney Marsh has always been a place of pilgrimage for naturalists. In 1980 dykes such as Puddledock Sewer and Jury's Gap, whose name commemorates the ancient office of the jurats, or drainage officers, of the marsh, were still supporting such rarities as the greater water parsnip[53] and the great silver water beetle.[54] The heart of the marsh, known as Walland Marsh, had often gone under water in winter, enticing large flocks of duck, and even the exotic spoonbill, which was a regular visitor in the nineteenth century. A spectacular return of birdlife followed the deliberate flooding of 1,000 acres in 1940 for military defence. The last big marsh flood was in 1960, when the romantic church of St Thomas à Becket at Fairfield was stranded like a dumpy ark in the midst of an immense inland sea. During and after the War, however, increasingly large areas were ploughed for wheat and potatoes. Romney Marsh was picked out for special mention by the wartime Agricultural Land Commission, and between 1939 and 1944 the proportion of land under the plough increased from 9 to 37 per cent.[55] By 1980 about two-thirds of the land south and east of the Royal Military Canal had been thoroughly drained, and the famous Romney Marsh sheep, the breed ancestral to all the flocks in the Falkland Islands, whose immaculate fleeciness had been admired by visitors ever since Cobbett wrote of them being 'as white as a piece of writing paper', were vanishing along with the waders and the yellow wagtails. The internal drainage boards, administered in this particular area by the Southern Water Authority, vigorously dredge and spray the dykes, which in many places are reduced to a soup of blanket weed, encouraged by nitrogen leaching off the ploughland. Until recently, the Southern Water Authority was even spraying the dykes of Romney marsh from the air. Elsewhere, ditches are filled in to make way for larger potato fields.

The very last refuges of surviving pasture and unspoiled dyke on Romney

Even the famous Romney Marsh sheep have become as scarce as the previously abundant wildlife on Romney Marsh. Photo: Jo Nelson.

Marsh have thus become increasingly precious. In 1982 the owner of 150 acres of wet pasture in Walland Marsh applied to the NCC for the compensation newly available under the terms of the Wildlife and Countryside Act, if he would forego the benefits of a drainage scheme which would enable him to convert his land to winter wheat. The pasture was within an SSSI and, besides having the remains of an unusual salt-marsh flora, currently supported breeding populations of snipe, garganey, bearded tit, and, rarest of all, black-tailed godwit.[56] The farm manager was sympathetic to the NCC, but he had waited two and half years for their decision concerning his application to the Ministry of Agriculture for a drainage grant. Conservationists all over the country waited to see what the council would do in this, the first test of the Act. The Department of the Environment refused to fund the council to secure a management agreement for Walland Marsh, and while the council had enough of its own money to pay the compensation required, it possessed insufficient spare funds to deal with the even more urgent cases certain to come up in the same year. In the last resort, the NCC, chief guardian of the nation's natural heritage, like Nelson fixing his telescope to his blind eye, turned its binoculars back to front, and, ignoring the nesting godwits on Walland Marsh, declared the site to be of no great natural interest. The land was de-scheduled as a special site, and the farmer proceeded

Black-tailed godwit.

to plough. The Royal Society for the Protection of Birds stated in the *Observer* that the NCC 'must object loudly when any site is at risk, whether or not it has the money to pay compensation'.[57] But it was scarcely in a position to protest too much, for, years before, the society had sold that very same land on Walland Marsh, thereby enabling it to buy its reserve on Dungeness.

The sequel to the saga of Walland Marsh came in 1984, when the drainage scheme initiated two years earlier on this, some of the lowest land in the entire marsh, proved quite inadequate to combat winter flooding. The land was put on the market, and the Kent Trust was approached to see if it would like to buy it back as a bird reserve. The trust had more urgent priorities to attend to than to buy the now birdless ploughland, and bad farmland at that. The affair of Romney Marsh was simply an opening shot in the wetland debate. The next major engagement was to come to a head almost exactly a year later, away in the West.

### The Somerset Levels

The Somerset Levels, between Wells and Taunton, do not extend to the horizon as one immense open space like Romney Marsh or the eastern Fens. The wetlands of Somerset consist of many fingers of damp grassland, known as 'moors', which are webbed between tongues of higher ground: the Mendips to the north, the Blackdown Hills to the south, and, between these, the Polden Hills and the long low ridges of the 'zoys', or islands, which give their names to such villages as Chedzoy and Westonzoyland. The villages cluster round the edges of West Sedgemoor, King's Sedgemoor, Tealham, and Tadham Moors, like lakeside settlements, which is indeed what they once were. Grey houses on the edge of a green moor, they are withdrawn, but only just, from the lowest sumps in what Adam Nicolson has well described as 'a *poured* landscape', held in place by the low corrugations of the drier ground.[58]

The Somerset Levels do not fit comfortably with our conventional notions of the West Country—cosy cottages and cream teas. There is even something other-worldly about the Levels' two most ever-present creatures, the presiding spirits of the place: the eel and the heron. A slightly forbidding quality characterizes

(a)

(b)

Typical flowers
of the wetlands.

(a) Flowering rush.

(b) Valerian.

(c) Kingcups.

(d) Cotton grass.

(e) Ragged robin.

(f) Yellow flag iris.

(g) Great spearwort.

(c)

(d)

(e)

(f)

(g)

Otmoor. Winter sunlight picks out the beacon of Charlton church, which used to guide lost travellers out of the wetland.

The eel and the heron are the guardian spirits of the Somerset Levels.

this countryside, as with all land where water lies waiting, just below the surface, to rise at flood-times, sealing off the central fastness of many of the moors, even from the silent cattle. The strangeness is emphasized by the isolated dome of Glastonbury Tor, crowned by the roofless beacon of the church where, in 1539, the last abbot of Glastonbury was hung, and by the Tor's smaller twin, the Burrow Mump, also bearing the stump of a ruined tower on its windy eminence. Wherever you go among the quiet pastures and beside the ditches, known here as 'rhynes', gelatinous with duckweed and bladderwort and shaded by lines of leaning pollards, the improbable silhouettes of the Tor or Burrow Mump draw the eye like lighthouses in a green sea, and seem to focus the odd, individual character of the whole region. Michael Williams, the historian of the Somerset Levels, has described one of those special moments when 'just after sunrise, the Levels are often covered with a thick mist, like a sea. Nothing rises above the flat white surface except the smooth Tor of Glastonbury surmounted by the slender shape of St Michael's Tower, looking for all the world like the Isle of Avalon it is claimed to be.'[59] At other times, from Glastonbury Tor, you can watch the procession of clouds blow in from the Atlantic, to spill their incessant rain upon the Levels. Although this part of Somerset is drier than some parts of the West Country, its mild moist climate ensures that everything grows and is grown upon. On the peatlands at Shapwick, the dewed luxuriance of royal fern competes with an abundance of bittersweet and bog myrtle. On the moors, pollard crowns sprout a beard of grass and tangled briars. A down of lichen grows over the willow trunks, and even on the gargoyles on St Michael's tower, and gold-green goitres of mistletoe hang in the branches of the cider-apple trees.

The saturated nature of this landscape, with so much water falling from above and oozing from below, is the key to understanding the special problems in Somerset, and why this area became the focal point for the entire wetlands issue in 1983. Although the whole region is generally known as the Somerset Levels, strictly speaking, the Levels are a clay belt of less outstanding environmental interest lying between the moors and the sea. This clay belt is significantly higher than the fingers of grazed moor which extend inland, and which collect all the water coming down from the upland to swell the rivers Brue, Parrett, and Axe. Compared to the Fens, where the ratio of upland to lowland is two to one, the ratio between the Somerset and Dorset hinterland and the lowland moors is four to one, so there is a great deal of water flowing down off the hills. Much of this water flows across the Levels within embanked rivers, whose water-level is higher than the adjacent land. However, the water which falls on the lowest land, which at high tide is 14 feet below sea-level, must be pumped *up* and out over the clay belt before it can reach the sea. Even then, the flood-water cannot escape into Bridgwater Bay if the tide is in. The Bristol Channel has a 40-foot tidal range, the second highest in the world, and so the water must be stored in the rivers, to be let out by gravity as the tide falls.

All these circumstances, together with the high local rainfall, are the makings

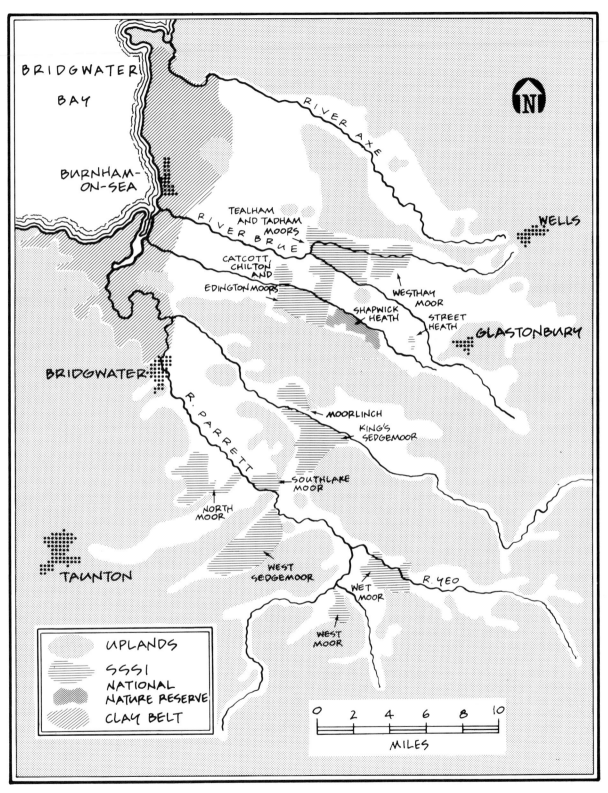

BRIDGWATER BAY

BURNHAM-ON-SEA

RIVER AXE

WELLS

TEALHAM AND TADHAM MOORS

RIVER BRUE

CATCOTT CHILTON AND

EDINGTON MOORS

SHAPWICK HEATH

WESTHAY MOOR

STREET HEATH

GLASTONBURY

BRIDGWATER

R. PARRETT

MOORLINCH

KING'S SEDGEMOOR

SOUTHLAKE MOOR

NORTH MOOR

WEST SEDGEMOOR

WET MOOR

R. YEO

WEST MOOR

TAUNTON

UPLANDS

SSSI

NATIONAL NATURE RESERVE

CLAY BELT

0  2  4  6  8  10
MILES

Somerset Levels.

of a drainer's nightmare or, at any rate, an engineer's interesting problem. It is not surprising that the sophisticated management of water has always shaped, and still dominates, life in the Somerset Levels. An awareness of the threat of rising water is firmly rooted in the local imagination. There is a common belief that the top of the tower of Mark church is level with the land at Burnham-on-Sea. This statistic does not bear close examination, but a woodcut of 1607 illustrates a baby in its crib bobbing by on the flood past the top of Chedzoy church tower. Such disasters have not been confined to the historic past. Adam Nicolson reports how the lock-keeper at Oath was woken by his wife on the night of the great flood of 1929: '"Hark at that water out there" she says to me. We could feel the vibrations, we could feel it humming up through the house. "That water's going over the top", she says. . . . She knew it was coming in her sleep.'[60]

In the flood of 1981, one farmer near Pawlett lost all but two of his 1,200 pigs. Sheep floated into houses and were marooned on the tops of hedges. The uncertainty surrounding an expansionist agriculture in such a place has helped dictate another striking feature of this landscape: the small size of the fields. The cider orchards are seldom larger than many people's gardens, and the pasture is parcelled into tiny plots, divided by a grid of rhynes, which act as wet fences between one small herd and the next. On West Sedgemoor, in an area which in eastern England might, typically, be shared between just two farmers, there were, in 1978, around a hundred owners and occupiers. On the Levels as a whole, the average farm size is 85 acres, each holding supporting around fifty cows.[61] For some farmers, land on higher ground provides their principal livelihood, to which the scraps of grazing on the moor add just a small supplementary income. Others are heirs to damp grassland which has been divided among the children of each succeeding generation. This fragmented land pattern, combined with the inaccessibility of many moors in winter, explains why so little change had occurred in the Somerset Levels by the 1970s, and consequently why there was so much at stake here, both for conservationists and would-be improvers.

The farming community, like the extraordinary community of flowers and birds which flourished on the farmland, had survived changes which have altered most of the rest of our countryside beyond recognition. Nearly half the family names associated with land ownership on West Sedgemoor in 1982 occur in the Enclosure Award for the same area for 1816.[62] Richard Cooke, who, from 1983, was employed as communicator between agriculture and conservation on the Levels, relates how he had to draw up family trees in the office, in order to work out the complex relationships in an area, where there could be up to thirty landowners with the same name. He also needed to know which branches of certain families were in a state of open hostility with their relatives. Frequently, on arriving to see a farmer in the afternoon, he would discover that his morning's negotiations on the neighbouring farm had been thoroughly aired by the entire community over a cider at lunch-time. Under these circumstances, rumour was always rife, and consensus among the farmers themselves on the merits of drainage was non-existent. A few more progressive farmers were in favour of large-scale drainage, and had even installed their own pumps. Some thought that they might like to drain, but they had never got around to it. Many had no desire to drain, but fiercely resented being told that they couldn't. And a good 20 per cent on West Sedgemoor had always believed that any drainage which resulted in 'breaking the spine'—the grass sward on the moor—would damage the land. In the words of a senior official in the Taunton office of the Ministry of Agriculture: 'If you asked a farmer on West Sedgemoor in 1975 about conservation, he probably wouldn't know the meaning of the word.' Yet, with no capital and minimum expenditure, these men were practising conservationists in the true meaning of the word. The fact that they were keepers of a landscape of whose national and international value they were still blissfully unaware was soon to change life on the Levels for all of them.

The ecological diversity which has made the Somerset Levels justly famous is remarkable, because it combines everything which a fine wetland should have: otters, wildfowl in winter, migrating birds in autumn and spring, ditches alive with all the dragonflies and flowers which a naturalist could hope to find, and, in summer, breeding waders and a tapestry of meadow flowers extending over the major part of the moors. The sheer scale of this abundance is exceptional. The part of the Somerset Levels regarded as of major interest extends over more than 100 square miles.[63] West Sedgemoor alone, much of which, before haymaking, is dense with ragged robin, green-winged orchids, marsh orchids, yellow rattle, and all the butterflies which come with such pasture, comprises approximately 2,500 acres. Compare this with a typical rural county such as Warwickshire, where only 100 acres of fine unimproved pasture survives, and that is scattered throughout the county. In winter, when Europe and eastern England are gripped by ice, the mild washlands in the West provide open water for internationally significant numbers of wild duck. In 1981–2, 30,000 wigeon and 4,800 teals were recorded, together with 250 Bewick's swans. Immense flocks of waders also arrive on the moors in winter. In the mid-1970s 8,000 dunlins and 2,000 golden plovers were counted on the Levels; while in 1980, 56,000 lapwings alighted on the green pastures. Short-eared owls, hen-harriers, and up to six peregrine falcons regularly quarter the open country around Glastonbury and Burrow Mump in the cold season. In spring, our largest population of whimbrel gather strength on the moors, *en route* for their breeding grounds in Iceland; and it is then that the whinchat, redshank, snipe, yellow wagtail, and curlew return to the Levels to breed. The rhynes are the home of our smallest flowering plant, *Wolffia arrhiza*, amongst all the juicy luxuriance of water violet and flowering rush. They also support such scarce beauties as the Hairy Dragonfly, *Brachytron pratense*, whose elegant black body is jewelled with blue-green markings, and the brilliant crimson Ruddy Darter Dragonfly, *Sympetrum sanguineum*. As if this wasn't enough, remnant raised mires show a spectacular, though increasingly excavated, peatland flora. The Somerset Levels are also a storehouse of archaeological treasures of international importance: Romano-British earthworks, Christian sites from the Dark Ages, sixteenth-century duck decoys, and timber causeways, including the famous Sweet Track, which may be the oldest artificial roadway in the world, dating back to around 4000 BC.[64]

As this ecological and archaeological heritage began to come under increasing threat from drainage, from 1962 onwards, the NCC set up a working party, which in 1977 published a consultation paper entitled 'The Somerset Wetlands Project'.[65] This document, which was circulated nationally, was the first to really alert people to the scale of the threats from agriculture and drainage, as well as to the environmental value of what was at stake. It also came up with the concept of identifying and protecting 'key' areas, where conservation should take priority over agricultural intensification. What also became clear was that such key areas could not be whittled down to just a few meadows here and there. Because of the interrelation of ecological systems and the integrated management of water throughout the Levels, certain moors had a total habitat value which was far greater than the sum of their parts. This ecological richness is, of course, the highly evolved product of a sophisticated manipulation of water-levels, carried out in the interests of dairy farming: the maintenance of high water-levels in summer for irrigation, the control of peak winter flooding, ensuring that the moors remain as pasture, rather than reverting to swamp, and the rhynes themselves, the key to the whole drainage system, whose controlled management guarantees a specialized diversity of wildlife. Drainage, which for so long had subtly sorted land from water into a fine-tuned harmony between man and nature, was now poised to destroy what it had previously created.

In 1976, Ralph Baker, who farmed some of the lowest-lying land in the Brue valley, exerted a powerful influence on the internal drainage boards, and was, above all, chairman of the regional land drainage committee of the Wessex Water Authority, published an article entitled 'Lincolnshire Fens in Somerset'. He reported on a trip taken by internal drainage board chairmen to visit the Fens, where they admired the comprehensive pumped drainage and the resulting abundance of wheat, potatoes, and sugar-beet.

We came back feeling hopeful . . . that Somerset will be drained and also aware of our responsibilities for this as members of Internal Drainage Boards in Somerset. . . . With the coming need for more food to avert food shortages and to lessen our country's balance of payments problems, the fuller utilisation of the Somerset Lowlands could become a project of prime importance with massive capital injection from national resources.[66]

The only really effective way to those national resources was via the Wessex Water Authority, for only it had the money to spend and the power to control the main rivers into which the water from farmers' pump-drained schemes ultimately emptied. In those early days, with Ralph Baker behind them, there was no lack of enthusiasm from the water authority to tackle the engineering challenge presented by the goal of full-scale drainage of the Somerset Levels. As one Wessex Water Authority report emphasized: 'The River Authority has tended to pioneer the way. . . . It requires the courage of the first few to prove a belief before the majority will join in and take advantage of our benefits.'[67]

The majority of farmers took a lot of convincing, however. The debate as to whether Somerset ever had the agricultural potential to become another Lincolnshire is not easily decided. It certainly would never have made good corn country, for, in addition to heavy rainfall, biting frosts can linger, on West Sedgemoor in particular, until as late as June. Within two years of a drainage scheme, one land-holding on West Sedgemoor suffered ochre problems, and the hot summer of 1984 proved the value of damp grassland on the moors at a time when surrounding dairy farmers were suffering from drought. On a sliding scale of 1 to 5, the Ministry of Agriculture has always listed the Somerset Levels as grade 2 on the basis of their potential productivity. However, on close examination of the guide-lines for classification, it has been found that, due to an error in not taking account of the enormous costs of drainage, this land should really have been classed only as grade 4. None the less, potatoes, celery, and especially carrots thrive on this rich peaty soil. One grower on the Levels produces carrots of almost legendary proportions: as thick as your wrist and said to last for four meals. It is true that if everyone converted to these crops, markets might prove hard to find; and general ploughing of the drier land would certainly trigger peat wastage. In reality, the major harvest of comprehensive drainage would probably be more and better grass—four crops of silage a year, and a tripling of the stock which the land could sustain.

In 1977 that possibility was presented to the farmers on West Sedgemoor. The water authority proposed a scheme, valued at up to £1 million, to upgrade the local pumping station and improve the main drain down the centre of the moor. The fate of the finest jewel in the Somerset Levels hung in the balance. After lengthy public deliberations, West Sedgemoor was, in the last resort, saved by the farmers themselves, who turned the scheme down. They feared an increase in rates, especially as a result of their anticipated dependence on piped water for cattle troughs once the water-level in the ditch system had been lowered, a change which would also necessitate expensive fencing. Another factor was probably that characteristic of wetland inhabitants, which led Gloucestershire farmers in the 1970s to oppose progressive drainage schemes for the Severn washlands, epitomized in the words of one puzzled official as 'an innate spirit of independence which seems to pervade the parish of Chaceley'.[68] By 1978, when the Royal Society for the Protection of Birds began to buy land on West Sedgemoor, it was too late for Somerset farmers to change their minds. The environmentalists had arrived, and as they began to drum up national opposition to the major drainage schemes proposed for the river Parrett and the river Brue by the Wessex Water Authority, they achieved something which had always eluded the officials of agriculture in dealing with this feuding rural community:

a virtually unanimous consensus among farmers of the Somerset Levels, in this case as they united against a common enemy, the conservationists.

In the Somerset issue there was a real dilemma, which is bound to surface repeatedly, since most of us, many conservationists included, subscribe to a life-style pinned to expectations of improvement, tied to an overall system of economic growth. Many farmers on the Levels are trapped, if they stop to think about it, by their low-input agriculture, which allows them no capital to escape from the parish in which they were born. It is fortunate for the environment and for the rest of us that generations of stock farmers and withy-growers have generally been content with their excellent cider, their close-knit community, and the level green horizons which contain their world. How many environmentalists, as they drive back up the motorway, having admired the birdlife of the Levels through a brand-new pair of binoculars, pause to consider that they are really having their cake and eating it too, as they enjoy a top-quality environment which may be preserved only because the quality of the lives of its inhabitants holds out little promise of advancement. This, at any rate, was one of the feelings current at the time, and there was no lack of students, authors, journalists, film-makers, and conference delegates, who gathered from cities and universities to study the ecological and social phenomena which constitute the Somerset Levels. One cartoon which went the rounds showed a huddle of bottoms gathered around some rare species on the moors being observed by a pair of swallows who commented: 'It is remarkable how they always return each year!' Adam Nicolson has recorded the common-sense exasperation which some local people felt about conservation:

and what's it done to the farmers and the drainage men? I'll tell you: we're the endangered species now. You should see us. We've all got copies of the Wildlife and Countryside Act 1981 at the bedside now, seeing if there is any mention in there of the Greater Crested Farmer or the Lesser Spotted Drainage Engineer. You look in any of these farmhouses round here at eleven at night and they'll all be reading a couple of chapters of the barmy thing before dropping off to sleep.[69]

The way in which people's lives become entangled in the struggle to defend an ecosystem validated in strictly scientific terms was starkly demonstrated by the drama which followed the NCC's 1983 decision to designate West Sedgemoor an SSSI under the terms of the new Act.

West Sedgemoor, shaped like a leaf and veined with gleaming ditches and haphazard lines of willows which delineate some of the many land-holdings on the moor, consists of about 4 square miles of damp grassland and withy-beds within a circle of low hills at the southernmost corner of the Levels. The tracks which lead on to the moor soon peter out on its soft surface, and in winter the pasture is largely inaccessible. The cattle are then taken off to be housed elsewhere, and in summer even silage making is impossible in the heart of the moor because of problems of access. In a few places on West Sedgemoor, salt-

West Sedgemoor was fiercely fought over by the naturalists and the drainage men.

tolerant plants still survive, as a ghostly reminder of the distant past, when these pastures were part of an estuarine lake. The inaccessibility of the place is part of its fascination. Viewed from the hill, it is like the garden which Alice in Wonderland tried to get into, but could never quite reach, and 'garden' is the right word to use for this extraordinary assemblage of flowers, birds, and insects. Of all the moors on the Somerset Levels, West Sedgemoor supports the finest combination of wintering and breeding birds and flowery rhynes and meadows. Richard Cooke recounts how, after many hours of patient work persuading farmers to accept management agreements on the moor, he would begin to wonder whether the place was really worth all this effort, when a thousand lapwings would suddenly fly down to settle on the fields. In 1983 West Sedgemoor supported flocks of 25,000 lapwing, as well as between a third and a quarter of the population of snipe, curlew, and redshank of the entire Levels.[70] In addition, West Sedgemoor stands out as a single hydrological and ecological unit.

For these reasons, it was an obvious first candidate for designation as an SSSI, and, in making this decision, the NCC had its back to the wall. After the debacle on Romney Marsh, voluntary conservation bodies were waiting to create an outcry if the council faltered in its duties; and, indeed, the Royal Society for the Protection of Birds had threatened to bring legal action against the council if it failed to designate West Sedgemoor and the Berwyn Mountains in Wales as special sites. On its other flank, the council faced an agricultural lobby which had little but contempt for it. Both sides knew that the success or failure of the attempt to save West Sedgemoor would also decide the fate of the wildlife and the restrictions put upon farmers on the other moors in the valleys of the Parrett and the Brue, including King's Sedgemoor, North Moor, and Southlake Moor. Not only that, but West Sedgemoor, together with the Berwyns, was the ultimate test case for the success or failure of the Wildlife and Countryside Act. Continuing damage to the moor also meant that conservationists had to act fast. In March 1982 the NCC sent all the farmers on West Sedgemoor a memorandum written in the language of an income-tax demand, which certainly gave support to the claim that conservationists might understand everything about the behaviour of water beetles, but still had a lot to learn about ordinary human beings. 'In accordance with sub-section 28(2) of the Wildlife and Countryside Act 1981, I write to advise you that the Nature Conservancy Council is considering notification of the area indicated on the attached map,' it began. The area in question was the land belonging to the recipients of this masterpiece of government jargon, which proceeded to list 'Potentially Damaging Operations', as including 'All forms of cultivation. . . . Changes in seasonal pattern of grazing. . . . Any tree or shrub planting', and 'Infilling of ponds and other water bodies'.[71] This is not the kind of communication to which Somerset farmers are especially accustomed. Some landowners on West Sedgemoor were later to spend as long as three days discussing everything from the weather to the cider crop with conservationists before they finally consented to management agreements on their land.

In 1982 and 1983, with barely more than two officers to cover the region, the council was quite unable to devote the kind of time required to make special site designation acceptable on West Sedgemoor. In addition, it had not been provided with the financial details which would have enabled it to make precise promises to landowners on what kind of money they would be offered in compensation for profits foregone under the terms of the Act. Such precise information, together with a real threat of compulsory purchase, was to make the task of later negotiators a great deal easier.

Eight months of wrangling, during which the agricultural lobby did everything in its power to dissuade the conservationists from declaring the whole of West Sedgemoor an SSSI, dragged on until November 1982, when the council finally decided that it had no option but to stick to its guns. Its decision to notify the entire site as an SSSI was greeted with surprise and relief by voluntary environmental bodies and outrage by agriculturalists. The president of the National Farmers' Union announced that he was 'extremely dismayed', and in Westminster, the Somerset MPs Edward Du Cann and John Peyton made personal pleas to the Secretary for the Environment to intervene. Sir Ralph Verney, chairman of the NCC had, they suggested, been 'quite unable to keep his zealots and minions in any kind of effective check'. At the time, it was easy to recall Coleridge's quip about the river Parrett, which flows past West Sedgemoor, of which he wrote in 1798 that it looked so swollen with mud as if 'all the parrots in the House of Commons had been washing their consciences in it'.[72]

Comments on the rights of landowners by the highest in the land were echoed out on the moor. In the pub garden of the Black Smock, a makeshift gallows was erected. From it were suspended effigies of leading conservationists, complete with binoculars made from old loo-rolls and the inevitable butterfly nets, those eternal symbols of the feather-brained naturalist. These effigies were accordingly burned, together with copies of the notorious letter, while an old army tank draped with a banner announcing 'Dictatorship Ends Here' was trained on the smouldering symbols of conservation. In January 1983, Tom King, MP for Bridgwater, was appointed Secretary for the Environment, and one of his first acts was to tell Sir Ralph Verney that he would not be reappointed as chairman of the NCC. At least, that is a polite way of putting it. Following this announcement, Richard North, then environmental correspondent for *The Times*, was informed that Sir Ralph had announced to the council that he had been sacked.[73] Accordingly, the national media were unequivocal in their statements that the boss of the Government's chief watch-dog for nature conservation had been sacked for doing his job.[74]

But West Sedgemoor was now protected by law, and the *Guardian* reported that, 'in starting what amounts to a campaign of civil disobedience, the farmers are risking fines of up to £1,000 and £100 a day for as long as they flout the rules'.[75] The next challenge facing everyone was to make the rules work. In March, Tom King addressed a highly charged private meeting on West Sedgemoor with all those concerned. He offered crucial reassurance on compensation to farmers, including whatever was necessary to make up for any possible fall in land values, due to limits put on agricultural expansion by the terms of the special site designation, together with the appointment of a liaison officer employed jointly by the NCC and the Ministry of Agriculture to negotiate management agreements with all landowners on West Sedgemoor. Both sides in the debate agreed that this meeting did much to set the new deal for the Somerset Levels on the right footing. As the Minister rightly remarked: 'We are right in the front line of battle to preserve the voluntary approach to conservation.'

Another factor which certainly saved the day for the Somerset Levels was the agricultural over-production crisis, which reached major proportions throughout the country at just that time. In 1986, hay barns on the Levels were stuffed full of the previous year's harvest of grass, that commodity whose massive increase

Effigies of the conservationists burned by Somerset farmers in 1983. Photo: Roger Hutchins.

had been one of the main objects of the proposed drainage schemes. At the same time as the storm broke over the Somerset Levels, a pitched battle was being fought over another internationally important wetland, at Bharatpur in India. In November 1982 an angry crowd of villagers marched on the reserve, which is a wintering ground for the nearly extinct Siberian crane. The people were seeking firewood, together with grazing for their cattle and buffalo. The authorities offered them fodder; but finally, in order to enforce national park law, the police fired on the crowd, killing six villagers.[76] This tragedy reflects an extremely difficult problem for conservationists in the Third World—a head-on conflict between hungry people and wildlife. How disgraceful, therefore, that we in Europe should still be arguing over the fate of our best wetlands, when nowhere could our farmers be said to be in serious discomfort, and in many places we are stockpiling food.

What is the future for the Somerset Levels now that the politicians and the television cameras have departed? The water authority, as a whole, has since taken a relatively neutral and honourable stance on the issue of drainage in Somerset, although the large farmers who dominate the regional land drainage committees, who set the whole problem off in the first place, remain in power. The Somerset County Council, which in its first plan for the Levels was cautious about asserting a prime place for conservation, has become more confident in criticizing expansionist agriculture.[77] In the long run, however, the attitude of the Ministry of Agriculture is that, should there ever be another food crisis, the agricultural resource of the Somerset Levels lies waiting to be drained and harvested, beneath the frail fledge of that other resource, the moist square miles of ragged robin and hay-rattle. Meanwhile, it will take a generation to learn how to operate the details of the new bureaucracy designed to promote traditionalism.

Local conservationists still believe that erosion of habitat continues on the Levels, though at a much slower rate than before—the death by 100 cuts, compared to what was previously death by 1,000 cuts. In 1983 drainage was slipped by on part of Tealham and Tadham moors under the now-reformed three-months loophole. In 1986 the internal drainage board, which had always deliberately flooded Southlake Moor in hard winters, made a decision, without the agreement of the NCC, to cease this practice. The result was the loss of one of the most important washlands for wildfowl in the West. In November 1987, however, the decision never to reflood Southlake Moor was reversed. Surveys by the RSPB completed in 1987 and 1988 indicate a severe decline in breeding waders, yellow wagtail, and whinchat on many parts of the Levels. This is linked to a reduction of spring and summer water-levels, which appear to result from manipulation by man rather than simply climatic change.[78]

None the less, a degree of drainage, or, to be precise, careful control of water-levels, remains essential on all the moors if the fine balance required for both stock farming and wildlife is to be maintained. Conservationists, as well as farmers, were worried about milk quotas in 1984, and indeed, this has inevitably led many stockmen to plough some of their pasture, as a cost-cutting exercise, in order to grow their own roots to feed their cattle. A nation-wide fall in land prices also leads to less attractive compensation payments from the NCC. To some it may seem that the arrangement whereby farmers are paid not to work their land in accordance with real economic demands is a kind of defeat, and that the landscape of the Levels is now going under dust-sheets. But at least our options regarding what we want from this place remain open. It is easy to imagine that, in time, bird-hides and controlled access to the meadows on West Sedgemoor will offer thousands of people the privilege of admiring its secrets; then the money which its exceptional wildlife will earn for the guardians of the moor will simply be accepted as another form of farm income.

### River Soar

In 1986 the Somerset Trust for Nature Conservation commented that its stance in criticizing one of the latest drainage proposals on the Levels, the Thorney Moors scheme, was casting conservationists in the role of defenders of public money, rather than wildlife. This was certainly a major part of the case in a campaign mounted against a large river engineering scheme in the Midlands in the summer following the show-down in Somerset.

The river Soar between Leicester and its confluence with the Trent near Nottingham winds lazily among beds of sweet flag and broad flotillas of yellow water-lily. It bears the dubious distinction of being the river into which were flung the unloved bones of Richard III after they were disinterred during the Reformation from a makeshift grave, where they had lain since his defeat at nearby Bosworth Field. The Soar is one of the widest rivers in that part of England, and every summer it is enjoyed by holiday-makers in boats and barges, as they take advantage of the navigation locks installed in the eighteenth century. The river has always flooded in winter, as the youthful Osbert Sitwell discovered when he was taken on a pilgrimage by his father to visit the tombs of his ancestors in the church at Ratcliffe-on-Soar. They found the building flooded out: 'the recumbent effigies of knights and their ladies seemed to float on a flat mirror of water. . . . My father refused to be depressed, and merely called to Robins, who was in attendance outside: "Robins, another time remember to put in my gum-boots".'[79] Such flood problems, exacerbated by the way in which navigation structures raise the water-level, presented possibilities for the Leicestershire-based engineers of the Severn–Trent Water Authority, which had long cast covetous eyes on the Soar. Accordingly, it proposed a substantial lowering of the river over the 15 or so miles of its lower reaches, at an estimated cost of £6.4 million. Although such schemes normally slip by without any planning permission or public inquiry, the presence of navigation on the Soar ensured that in 1983 the Severn–Trent Water Authority had to go through the legal technicality of promoting its own private bill in Parliament to gain approval for the scheme.

This was an opportunity for which conservationists had been waiting ever since the Amberley inquiry in 1978. The door was now ajar for them not only to publicly question the specific merits of the Leicestershire scheme, but also, more important, to resurrect the debate over the questionable economics of agricultural land drainage, together with the secrecy vigilantly maintained by water authorities concerning their cost–benefit calculations. Environmentalists accordingly presented two objections to the river Soar scheme. Their first objection was that the resulting flood protection would lure farmers into switching to arable, or else intensifying their grazing, and thereby remove the matrix of hedges, old ash trees, and little ponds brimmed with crowfoot which make the Soar valley one of the finest corners of Leicestershire. No important wetlands were at stake, since existing SSSIs, such as the Lockington Marshes, were to be carefully retained through a system of sluices controlled by the NCC. Their objection was that the water authority had stitched together an economic package in which the problems of an occasionally flooded building here and a flood-prone road there had been used to artificially raise the otherwise marginal cost benefit of a scheme chiefly aimed at benefiting farmers in a way which could only increase our burgeoning food mountain.[80]

The authority maintained, correctly, that in its calculations, it was simply following orders from central government, in the form of Ministry of Agriculture ground rules for cost benefit. One way to change ground rules, however, is to establish a political precedent, and it was for precisely this reason that environmentalists set out to make a test case of the river Soar. Their success in changing the rules would depend upon the speed and deftness with which they used the slender opportunities presented to them by the procedure for a parliamentary bill. In the event, the political game which was played out between the protagonists of agriculture and the environmentalists was one in which the conservationists were always just one step behind their opponents.

Before proceeding with its scheme, the Severn–Trent Water Authority submitted feasibility proposals to the NCC, which then heard nothing until the bill was safely deposited in Parliament. The Royal Society for the Protection of Birds, which recognized the economic significance of the Soar Bill, did not hear from the water authority at all until the bill was in Parliament. One local MP raised some objections concerning potential damage to the wildlife of the Soar valley, however, and as a result, some safeguards were agreed upon. When the bill had actually passed into the Lords, the Council for the Protection of Rural England, having recognized the landscape implications of the scheme at the eleventh hour, persuaded Lord Beaumont of Whitley to set in motion a little-used parliamentary procedure to halt passage of the bill by calling a House of Lords committee to examine the evidence for and against the river Soar scheme. For the first time since Amberley, the economic validity of large-scale river engineering in the countryside was to be tested impartially.

On a sweltering July day, the Severn–Trent Water Authority presented its evidence in one of the great painted chambers which face on to the Thames at Westminster. The electric atmosphere of imminent thunderstorms outside was heightened by the drama inside the House, where, by coincidence, the controversial debate on whether to bring back hanging was being fought out in the Commons on the same day. In the Lords' committee room both sides waited for a real challenge to the rationale of land drainage to begin when the economists employed by the environmental lobby put their case. But they were never called to speak. After a short time, the procedure already having been stretched to the utmost, their lordships decided that they had heard enough, and the case against the Soar scheme was dismissed without being given a verbal hearing. It transpired that the Council for the Protection of Rural England had failed to formally petition against the bill before the closing date. Divided among themselves, and having missed the crucial dates at every stage in the bill, the conservationists had done well to get as far as a House of Lords committee at all. Under these circumstances, the whole procedure revealed not so much the might of the agricultural lobby as the growing power of the environmentalists, who had managed to bend parliamentary rules as far as they did.

Three years later, with large-scale engineering on the river Soar well under way, it was easier to judge the rights and wrongs of the particular case. In 1986 the first stage of the Soar scheme had not proved an environmental disaster. Copses, hedges, and an old withy-bed had been retained and extended; ponds had been created; and wide new bays had been shaped into the river bank. Despite this, however, the whole enterprise could now be seen more clearly than ever as a colossal waste of public money. The expense of patching bank erosion set in train by the scheme, together with compensation payments to landowners for disruption, were turning out to be costlier than envisaged, while gravel winning in the valley looked an increasingly lucrative alternative to growing grain. Most farmers and householders on the reach affected seemed more concerned about the temporary disturbance created by the engineering, than enthusiastic about its benefits. One of the few really large grass fields below Kegworth, where conversion to grain was advocated by its owner, belonged, by predictable coincidence, to the chairman of the local IDB, who also happened to sit on the water authority's local drainage committee. Yet, in 1986, despite government pronouncements that drainage for agriculture should cease to be given priority, the land drainage committee of the Severn–Trent Water Authority pressed on with its proposals to continue to work up the valley into the 1990s.

*Otmoor*

In 1984 the Thames Water Authority proposed a £1.6-million drainage scheme to lower the river Cherwell in Oxfordshire. This would improve the outfall of its tributary the Ray, run off into which, from development at Banbury and

Bicester, was worsening the drainage on Otmoor, that most celebrated of battle-grounds over nineteenth-century drainage and enclosure. This time, however, those who opposed drainage on the moor wielded far greater influence than the men of the seven towns when they hurled sticks and stones at the yeomanry of Oxford in 1830. *The Times* reported that the water authority proposals had raised an outcry from Oxford's 'numerous and articulate amenity and con-servation groups', while one Oxford city councillor announced: 'We will unite all groups in the city to stop this appalling plan.'[81] Two years before, those who objected to a suggested M40 route across Otmoor had resorted to the ingenious strategem of buying numerous tiny parcels of land, whose ownership would then take the protagonists of the scheme a long while to trace: a delaying tactic against central authority of which the commoners of Otmoor would surely have approved.

In reality, the ecological value of Otmoor has already been much reduced. One large area has long been ploughed in the wake of an independent pump-drainage scheme; and the main refuge for such wetland specialities as marsh stitchwort and the marsh fritillary butterfly has been by and large reduced to the rifle range at the northern end of the moor. None the less, the issue concerned more intangible environmental qualities than simply lists of species. The thorn hedges on Otmoor, which enclose pastures with such names as The Goosey, Mousey's Corner, The Old Horse Ground, The Flits, still designated on the 1950 Ordnance Survey, and which had been the hated symbols of enclosure in 1830, were now part of a landscape which seemed to represent all the brooding atmosphere of our vanishing wetlands. Even now, when the westerly light sweeps over the empty moor on a winter's day, it never fails to spotlight the beacon of Charlton church, which in the past always guided lost travellers out of this hostile fastness. In 1984 the Thames Water Authority withdrew its drainage proposals for Otmoor, not least because of objections from those who feared that the scheme would also ruin the pleasures of punting on the Cherwell. The following year, the power of the environmental lobby was confirmed when an ambitious drainage scheme for the East Anglian Stour in Constable country was turned down on landscape grounds.

## The Derwent Ings

The autumn of 1984 also saw the finale of a long-standing battle over drainage of a river valley which is less famous than Dedham Vale or Otmoor, but of greater ecological importance than either of them, and also one of the wildest and most spectacular wetlands in England. In 1540 Henry VIII's chaplain, John Leland, wrote of the Yorkshire Derwent: 'This ryver at Greate raynes ragith and overfloweth much into low meadowes.'[82] These meadows, extending alongside the river for 12 miles or so south of Sutton-upon-Derwent, and known as the Derwent Ings, still go under in most winters, and when they do, the sense of an

A winter flood on the Derwent Ings, Yorkshire. Photo: Glyn Satterley.

untamed flood over the Ings is formidable. One part of the valley, the Wheldrake Ings, takes its name from the Anglo-Saxon word 'cweld', meaning death. Whether this referred to destruction by flooding or simply to the drowning of felons in the Derwent is uncertain. When snow melts on the north Yorkshire moors, it swells the racing, icy waters of this, one of the cleanest of all our lowland rivers, and so drowns the entire sweep of the valley, flowing right up towards the graveyard at East Cottingwith, with its sentinel tombstone of Snowden Slights, the legendary wildfowler, and lapping the churchyard wall at Aughton, most solitary of all wetland churches. Then the whole of the Ings, for a mile across, are given over to gleaming water and wild birds, the whistling of wigeons, and the melancholy fluting of the Bewick's swans.

But this great inundation is not hostile to the interests of farming any more than it is to wildlife. For centuries local landowners have known how to take advantage of the rich silt which the Derwent deposits every winter on their fields,

The full extent of a Derwent flood. Aughton church stands at the edge of the flood-water on the right. Photo: Cambridge University Collection of Air Photographs.

which creates, without any need for further fertilizer, some of the richest—and floweriest—haymeadows in Yorkshire. Here redshank and oyster-catcher breed among swards of marsh orchid, meadowsweet, and sweet vernal grass, which gives a special fragrance to the hay. At haymaking, this is gathered into farms built of mellow brick and pantile, typical of the East Riding, and then on Ings breaking day, the cattle are let in to graze on the open meadows. In May, snipe regularly perch on the turrets of the little church tower at Aughton, like wetland weathercocks, to perform their brisk territorial calls, known as 'chipping'. The church contains reminders of Robert Aske, who in 1536 set out from this lonely place to begin the rebellion which was to lead to Aske's own execution in York. Carved inside the tower are the words: CHRISTOPHER. SON OF ROBERT ASKE. REMEMBER 1536. Within the simple stone chancel, you can hear the curlew calling outside in the empty fields.

In 1975 the seasonal flooding which had always worked so well for the farmers

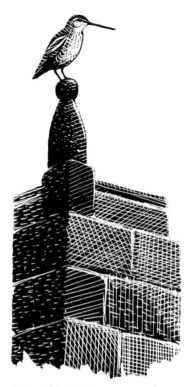

Snipe 'chipping' on a turret of
Aughton church, Derwent Ings.

on the Ings and had earned the 2,000-acre washland an SSSI status of international importance for its waders and wildfowl was disrupted. The Yorkshire Water Authority built a barrage across the mouth of the Derwent at Barmby to extract drinking-water. This scheme had been hurried through, following the public outcry which had thwarted proposals to build a reservoir at Farndale, and there had been little or no study of the possible environmental effects of the Barmby barrage. Once installed, however, the barrage appeared to some to impound the flood-waters upstream. Others blamed climatic change and the problems of runoff. The subtle management of the Derwent Ings, whereby flood-water was let off the land to flow back into the river through a system of carefully maintained ditches and flap-valves known as 'cloughs', ceased to operate effectively. At least, this is what some farmers believed, but the NCC insists that there is no quantitative data to support the claim that the Ings have become wetter since the construction of the Barmby barrage.

For almost ten years, the protagonists and the opponents of drainage were to argue over how to restore the status quo, but they could never agree on exactly how this should be done.[83] In 1980 the Ouse and Derwent Internal Drainage Board proposed a pump-drainage scheme for that part of the Derwent Ings known as the North Duffield Carrs. The board insisted that it wished only to remove the surplus flood-water which had affected the lower Derwent valley since 1975. Conservationists feared that, with another internal drainage board on the opposite bank of the river also eager for drainage, the North Duffield scheme would simply open the way for piecemeal ploughing of pasture, as was happening to wetlands all over the country at that time, and indeed near at hand, where the breeding grounds for snipe and lapwing were ploughed and sown with rye-grass on the Melbourne Ings beside the Pocklington canal. In 1982 the NCC provisionally agreed on a compromise with the IDB, concerning pump-drainage on the Derwent, but later withdrew its support when the National Farmers' Union would not agree to a 5-year monitoring of the scheme, together with abandonment of pumping should it prove detrimental to wildlife. The Yorkshire Wildlife Trust maintained that if the cost benefit for this £55,000 scheme were valid, then it must be assuming appreciable undeclared agricultural benefits. Either the proposals were economically unsound, or they were incompatible with wildlife. The shadow of drainage which hung over the Derwent was a very real one, especially since the first pumping proposals were submitted well before the

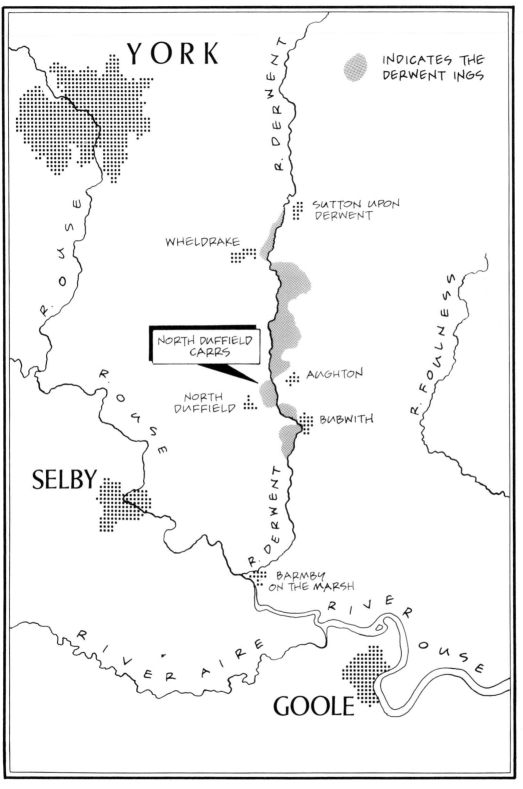

YORK

R. DERWENT

INDICATES THE
DERWENT INGS

R. OUSE

SUTTON UPON
DERWENT

WHELDRAKE

R. OUSE

NORTH DUFFIELD
CARRS

NORTH
DUFFIELD

AUGHTON

BUBWITH

R. FOULNESS

R. DERWENT

SELBY

BARMBY
ON THE MARSH

RIVER OUSE

RIVER AIRE

GOOLE

Derwent Ings.

1981 Wildlife and Countryside Act. In 1983 the IDB submitted its proposals to the Ministry of Agriculture for grant aid, and the Yorkshire Water Authority resolved to contribute the remaining 50 per cent of the cost of the scheme, should it gain approval. Conservationists waited to hear how, in the absence of any inquiry, the Minister of Agriculture would decide the fate of the Derwent Ings. At Christmas 1984, Lord Belstead, speaking for the Government at the annual dinner of all IDBs, beneath the chandeliers and caryatids of the Connaught Rooms in London, formally turned the scheme down. It seemed a long way from the curlews and the church at Aughton.

The IDB continues to maintain, with some justice, that land beside the Derwent is too wet. But the NCC is steadily buying up the contentious acres of North Duffield Carrs, where any wetland management will be strictly controlled by conservationists, and on their own terms. Had the earlier proposals been approved, there would have been uncertainty on both sides as to exactly how the effects on the habitat of the Ings could have been controlled. The NCC wrote of the anticipated effects of pumping on the Derwent: 'Ecology is not an exact science since the biological systems with which we are dealing are so complex';[84] while the IDB had used exactly the same expression to describe the lack of absolute certainty with which precise rates of pumping could be gauged. Initiating the drainage of a wetland is like opening Pandora's box. Once begun, the consequences of lowering water-levels can slip out of everyone's control. This was the case, not only in terms of habitat, but also in terms of farming practice, concerning the management of a wetland which became the centre of a debate in 1984. This debate which was to have political consequences for the rest of the English landscape concerned the Norfolk marshes, which soon became notorious among environmentalists as Halvergate.

### Halvergate

That area which has come to be known, collectively, as the Halvergate marshes, ever since the controversies surrounding its future, is the most easterly part of the Norfolk Broads. It consists of more than 6 square miles of uninterrupted and largely inaccessible grazing marsh, extending between the low hills and the busy boating broads of the hinterland and the open sea. Of all the wetlands in England and Wales, the Halvergate marshes impress the visitor most by their sheer immensity. As you approach them from the land, by way of the lofty flint church of Wickhampton, their great stillness extends as far as the eye can see to the dim smudge which is Great Yarmouth on the furthest horizon. Coming down to the marshes from the opposite end, past the Roman fort at Burgh Castle or through the Waveney Forest, which smells of dew and pine resin and echoes with the muffled cough of the pheasant, the thickets fall back at the edge of the marsh, and away in the distance you can just make out the low circle of the western hills. Spread out between lies mile upon mile of buff-green marshland, flecked

The wide empty spaces of the Halvergate marshes, Norfolk.

with the glitter of the distant Friesian herds, each beast reduced to the smallest mote on a level sea of grass. There is a sense here that, not only the land, but also everything to do with our busy, complex civilization, has withdrawn from all sides of the marsh, leaving it open only to the sweeping Everests of cloud and the moving shafts of light. Out over the marshes, flocks of starlings rise and fall, floating like shoals of fish in an ocean of emptiness, and the hare races in circles among the hooves of the quiet cattle. Halvergate is a vast and spacious arena, one of the last of its kind in England.

Guarding the edges of this great natural theatre, stand windmills, coloured black and white, like those other presences of Halvergate, the cattle and the lapwings. The seventy windmills of Halvergate are to the Norfolk marshes what churches are to most other wetlands, sentinels rising from the otherwise absolute level. Some are black and broken, robbed of their sails. Others are restored, their caps jauntily picked out in navy blue and white. On a bright day, they glint, one

Hare and cattle, Halvergate.

beyond the next on the far horizons; and sometimes, when a sea fog rolls out over the marsh, only the tops of the windmills rise above the mist. This is not a landscape to admire as if it were a composed picture, like some hill country. It is a landscape to be in, especially at its innermost heart of Haddiscoe Island, where there is nothing louder than the brushing of the reeds as they lean against each other, and nothing taller than the lichened gateposts which mark the crossings of the winding creeks. Here you are surrounded by an undiminished sense of openness and tranquillity.

The other special quality of Haddiscoe is that it remains a working landscape, grazed by stock farmers who are happy to continue this tradition. The way in which these men who make a living from the marshes also love the loneliness and the particular character of this place comes across very clearly in Billy Lacey, one of the five marshmen still employed to tend the stock on Halvergate. Of all the times he relishes on the marshes, where he has spent his entire life, he describes the best as being out in the absolute solitude of a frosty midnight beneath a full moon at lambing time. His sympathy with the place and its traditions, which is shared by many stock farmers, is by no means common to all who wield influence in the Norfolk Broads, however.

During the 1980s the immense arena of Halvergate has been the setting for a real human drama, in which the conflict is not so much between agriculture and the environment as between two different kinds of farmer. In this battle, small stockmen have been struggling for survival, as they risked being driven out of business by a few large arable farmers and progressive dairy farmers, whose survival was never at risk, but whose profits would be enormously increased by drainage and subsequent ploughing of the Halvergate marshes, which would allow them to grow corn or short-term grass leys for silage. The mechanism by which this take-over could be achieved lies at the very heart of the political structure whereby drainage is administered, especially the internal drainage boards.

What was at stake at Halvergate was no great wetland of prime ecological significance. Covering a relatively small area at the margins of the marsh is an SSSI, scheduled on account of the plants and dragonflies in its spring-fed ditches.

The main central area of grassland has long been drained to the point where conditions no longer encourage many breeding waders or winter wildfowl. Even in the great floods of 1953, not all the Halvergate marshes went under water. The environmental issue there concerned landscape, rather than habitat loss, and consequently the NCC played little part in the debate. The chief protagonists of conservation at Halvergate were bodies with a specific remit to protect the landscape: the Countryside Commission and the Broads Authority, an organization set up in 1978 'to conserve and enhance natural beauty and amenity' in the Norfolk Broads[85]—the total region, of course, of which the Halvergate marshes are only one part.

Even the case of landscape quality diminished as a result of drainage and ploughing of the marsh can be a hard one to argue. If Halvergate grew nothing but grain, there would still be the wide open spaces and the great empty skies. The Broads Authority, in its analysis of the landscape problems, was careful not to talk of such subjective criteria as beauty, but emphasized instead that ploughing would transform the character of the marshes from one dominated by 'rough texture, vegetated dykes, field gates, grazing animals and birds'.[86] The landscape, which remained substantially the same as that which had inspired the Norwich school of artists would then become a single field of wheat, like all the other East Anglian ploughland which surrounded it. These changes to such a special part of Norfolk would be part of the nation-wide erosion, through intensive agriculture, of that variety of local character which has always been the particular hallmark of the English landscape.

Behind the environmental issue in the Broads, however, lay a far wider political issue of national significance. In March 1984, a *Times* leader described the Halvergate marshes as 'the Flanders of the great war between farming interests and the objectives of nature conservation'.[87] The critical issues in this war were the expense to the nation, in terms both of draining and ploughing marshland, financed by the Ministry of Agriculture, and of compensation to those farmers denied drainage, financed by the Department of the Environment. What was further at stake was the workability of voluntary agreements between farming and conservation under the terms of the 1981 Wildlife and Countryside Act, together with the threat, if these failed, of the first steps towards real planning controls in the countryside. Halvergate was important not because there was a precious landscape at risk, but because the political controversy surrounding it grew until it began to threaten the whole edifice of the 1981 Act. Consequently it had to be saved, irrespective of its intrinsic value.

In the early 1980s, on the 7,000 or so acres of the Halvergate marshes, there were around 100 farmers, over half of whom owned less than 25 acres. The vast majority of the smaller farmers grazed stock on the pasture, and, with low incomes and low output, depended on drainage rates being kept low and water-levels fairly high to ensure good summer grazing on this dry, easterly side of England. Many marshland grazing plots had been left by Norfolk farmers to their daughters, to provide them with a small nest-egg in the form of the grass keep. This fragmented ownership resembles that on West Sedgemoor; but the big difference between Somerset and East Anglia is that most of the West Country is pasture, whereas the marshes of the Norfolk Broads are small islands of grazing land within the most productive and lucrative granary in the country. David Brewster of the Broads Authority has emphasized: 'Our criticisms have never been directed at the farmers out on the marshes, who have come to us, explaining that they wish to plough. They are simply being pressured by ridiculous economic circumstances.'

These circumstances have been dictated by the economic incentives of the Common Agricultural Policy at the national level and political pressure from internal drainage boards at the local level, where, in the early 1980s, the controls lay in the hands of a minority of arable farmers and large dairy farmers, who, with voting strength proportional to land owned, were able to control the balance of power on the board. Power in a place like Halvergate requires control of the balance of water-levels. This is the key to the manipulation not only of the landscape, but also of the entire rural community.

For a farmer who owned some grazing marsh as a supplement to his large cornfields on the hill, and who was encouraged to intensify by massive incentives offered for cereals in preference to livestock by the Common Agricultural Policy, together with major funding of drainage by the Ministry of Agriculture, the temptations were to prove irresistible. Such farmers would already possess both the combine-harvesters and the capital necessary to switch poor grazing land over to cereal production. They would also have the money to contribute their part of the massive increase in drainage rates levied by the local board. All they needed was control of the board itself, and that of course, even where board chairmen are sympathetic to the environment, is exactly what the largest farmers always do have.

Those who sit on the IDB are able to levy increases in the drainage rate on both occupiers and owners, in order to pay back the expenses of the drainage schemes which they instigate. They also have complete discretion over the period during which that debt must be repaid. Thus, instead of spreading the load over a number of years, they can massively increase the rate which a small stock farmer may have to pay in a year. Furthermore, there is no proper differential rating system—the arrangements in some cases have not been revised for 80 years—to ensure that the man on the poorest land pays proportionally lower rates.

The benefits which a small stock farmer will receive from drainage improvements for which he is forced to pay are minimal, and sometimes his situation is worsened. If you go to the partially drained pastures on the north side of the Halvergate marshes now, the first things you notice are the thistles. Thistles do not thrive on undrained permanent pasture, but on permanent pasture which has been subject to drainage; and if an owner cannot afford to switch to high-gear grass leys for silage or else to corn, they flourish with a vengeance. The white floss of thistle-down can be seen floating over fields such as these all over England. They are the true harvest and symbol of a wasteful, unjust national policy for drainage. In addition, a stock farmer under such circumstances will be faced with less water in his ditches, so his cattle may escape into his neighbours' fields; it will also be more difficult for them to find clean ditch-water to drink. On coastal marshes, arterial drainage may encourage salt water to penetrate the ditch system, and on some of the peatlands, ochre released into the ditch system through deep drainage compounds the problems. When a grazing farmer takes his stock off the land to be housed indoors for the winter, as is normal, he derives no benefit from the expensive pumping which keeps his neighbours' ploughland dry through the winter months. Under such circumstances, the grazier who can afford the change-over will find himself pressurized into converting to arable. The smallest farmer, however, faced with rate increases and the withering thistle-infested sward which was once good grazing and which he may no longer be able to afford to replace with expensive rotations of rye-grass, is likely to sell up. This is the kind of man who, in the cost–benefit appraisals of publicly funded drainage schemes, is described as a 'laggard'. His land will probably be bought by one of the large arable farmers who promoted the drainage scheme, as is tacitly admitted by the Lower Bure IDB's report on one of the most controversial drainage schemes on the Halvergate marshes, where it stated that part of the land

is in the ownership of a large number of people. It is usual when improvements are carried out that merging of interests occur, and that there would be an amalgamation and rationalisation of field shapes and size to increase the size to between 5 and 12 hectares for efficient arable cultivation.[88]

Professor Timothy O'Riordan, who, as a member of the Broads Authority, was in a position to know, has made the point: 'Certainly some Lower Bure IDB members were poised to make a killing by buying land, thereby benefiting from the estimated £4,000 per hectare taxpayer subsidy . . . available to assist them in growing a surplus crop.'[89] We can see clearly here that the old motto 'By Farmers. For Farmers' can be rewritten 'By Some Farmers. For Some Farmers. And Against the Rest'. As a result of these developments, grazing farmers on the Halvergate marshes were so desperate that in 1984 they canvassed a leading activist in the local group of Friends of the Earth as to whether he would be prepared to be nominated by them as their candidate for the IDB; and in 1986 another grazing farmer was seriously considering suing the board for the drying-out of good grazing land.

These have been the socially divisive consequences of what Professor O'Riordan has described as '"lily-gilded" drainage'. 'To save Halvergate', he has written, 'would be to save the nation from wasting its money and to bring politicians to their senses over how to marry agricultural and conservation

After drainage, thistles extend across the Halvergate marshes on those fields whose owners cannot afford to plough and re-sow with grass leys or corn. They are the true harvest and symbol of a wasteful and inequable drainage policy.

policies ... at least cost to the taxpayer.'[90] As things developed, the burden on the taxpayer promised to increase. In 1984 Lords Buxton and Onslow wrote to *The Times* that farmers were threatening to plough, and consequently: 'With 5000 acres at risk, compensation payments for conservation under the Wildlife and Countryside Act arrangements might cost the public purse ultimately as much as £1 million a year, index-linked.'[91] A major reason why this compensation from the Department of the Environment was being pushed up was the Ministry of Agriculture's willingness to fund and encourage drainage schemes on the marshes. Behind the struggle in Norfolk lay just another episode in what the *New Scientist* described as 'a tug-of-war between two government departments',[92] those two old enemies and master chess-players, the Ministry of Agriculture and the Department of the Environment, out of whose feuding the totally unsatisfactory arrangements for the administration of land drainage had been arrived at back in 1974. Expense aside, by 1984 the voluntary agreements between high-gear agriculture and conservation were patently failing to work. Since 1982, 1,009 acres of grazing marshes had been ploughed, some 873 acres without notice to the Broads Authority.[93]

It is against this background that the events which unfolded on the Halvergate marshes between 1980 and 1984 must be understood. They can be seen as a progress in which drainage proposals, first by the water authority, then by the internal drainage board, and finally by private farmers, were successively thwarted by the environmental lobby. But triumph for the environmentalists was only partial in all cases. The Yare barrier proposed by the Anglian Water Authority, the ultimate dream of drainage men in East Norfolk, has not been dropped altogether, only shelved. The board managed to push at least one of its proposed schemes through, while individual farmers, as we have seen, got away with a fair degree of ploughing. Professor Edmund Penning-Rowsell has written of the Halvergate saga: 'The conservationists' "victory" in fact leaves the factors encouraging drainage unaltered and the political power structure virtually unchanged.'[94]

Ever since the floods of 1953, engineers have advocated building a barrier across the mouth of the river Yare to reduce the threat of tidal surge flooding, which occurred both then and again in 1978. In 1979 and 1980 schemes estimated at costing up to £17.5 million were carefully considered and finally, but not irrevocably, turned down.[95] While such a major scheme would certainly have hastened the draining of Norfolk's marshes, it was still possible to carry out plenty of marshland drainage without it. In 1979 approval, together with a Ministry of Agriculture grant, was awarded to the Muckfleet and South Flegg IDB to carry out a scheme on the river Muckfleet. The result was canalization of half a mile of the river, which in turn promoted the conversion of thousands of acres of grazing marsh into arable land, thereby driving a number of grazing farmers north of the river Bure out of business.

In 1980 the six pumps which drained the river Bure and the river Yare were becoming defunct, and this gave the Lower Bure IDB[96] an opportunity to submit a £2.33-million scheme to the Ministry of Agriculture as eligible for a £1-million grant from the taxpayer.[97] The total package comprised a three-pronged attack on the marshes, consisting of the Seven Mile/Berney scheme, the Manor House scheme, and the Tunstall/Acle scheme. These schemes proposed replacing the old diesel pumps by electric pumps, and seemed set to achieve effective draining and ploughing of the Halvergate marshes. The Countryside Commission called for a public inquiry on these proposals, to which the IDB responded that it would not appear at such an inquiry and would finance its own scheme if denied a Ministry of Agriculture grant. In January 1981 the Countryside Commission, crumpling under pressure from the agricultural lobby, withdrew its request for an inquiry, although in 1982 it again called for a public airing of the issue.

In February 1982 the board unilaterally decided to drop the sluice on the main Fleet drain, thereby effectively lowering water-levels over the area which would have been drained by the Manor House scheme. This ensured that pasture adjacent to the Fleet could be converted to arable and that farmers on the SSSI could use the real threat of ploughing as a way of claiming higher compensation from the NCC—all without the inconvenience of pushing through the contentious pumping scheme. In the summer of 1982 the thistles of Halvergate advanced across the pastures of the central marsh.

In November of that year, the ministers of Agriculture and the Environment agreed on a compromise, without the embarrassment of a public inquiry, to settle the matter of the IDB schemes once and for all.[98] They supported the Tunstall/Acle Scheme, which could now go through with a 50 per cent grant, but turned down all proposals to increase the capacity of the pumps for the Seven Mile/Berney scheme. None the less, when the replacement of the Seven Mile pump was installed as part of the compromise, both timing and setting of the sill were controlled by the local board, thereby facilitating an approximate 6-inch lowering of water-levels. In addition, where access down a muddy track had always been sufficient, the Ministry of Agriculture was empowered to provide half the £180,000 cost of a concrete road leading to the new pump.[99] The remaining 50 per cent would have to be paid back out of the drainage rate, borne in part by many small graziers who gained no benefit from it whatsoever. It was thought that the real purpose of this road was not to achieve access to the pump, but to pave the way for combine-harvesters.

The Government's decision to turn down part of the IDB scheme was an important precedent for the first moves towards planning restraint on agricultural practice. But although the board's proposals had been settled, the problem of compensation payments was not going to go away. These continued to rise astronomically, absorbing an estimated one-third of the Broads Authority's expenditure; and when the authority tried to hold the cost down, some farmers threatened to plough unless they received what they deemed to be sufficient. In

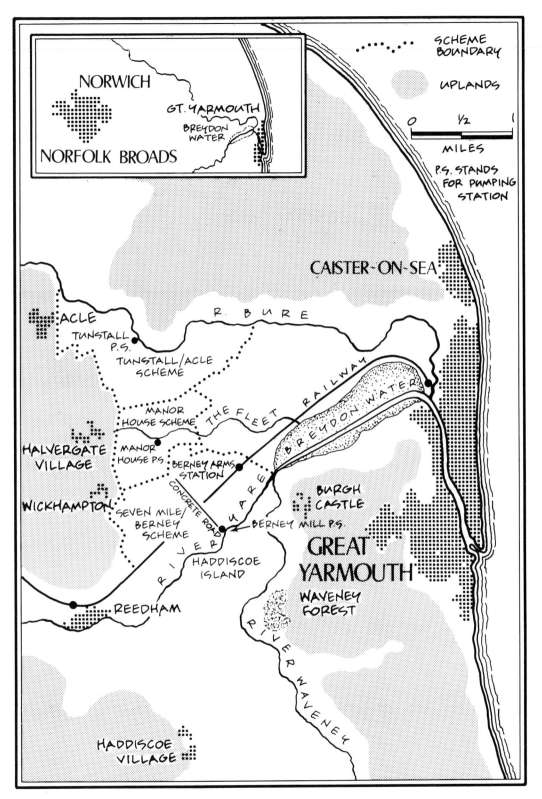

The Halvergate marshes.

In the early 1980s more efficient pumps began to arrive on the Halvergate marshes. Photo: Richard Dyer.

March 1984 government ministers agreed to look at the possibility of alternative financing as long as farmers would suspend operations for a year for a modest sum. This seemed satisfactory, so in April, William Waldegrave announced in the Commons that 'Halvergate was safe for a year.'[100] With so public a commitment, representing the Government's national stance on voluntary agreements between farming and conservation, all eyes were turned on Norfolk to see whether consensus in the countryside would really work.

In June, one farmer, dissatisfied with his level of compensation, began to clear out his ditches in preparation for his declared intention to plough. The Friends of the Earth immediately moved in and sat on the contractor's equipment. A ploughed field of itself is never news, but local activists seated patiently in the stationary dredger's bucket ensured coverage by the national media, who were driven out to the scene across the heart of the marsh. An old white horse was led in to join the motley gathering of local people and environmentalists. Startled by all this publicity, the farmer withdrew. In the same month, the Department of the Environment and the Ministry of Agriculture were unable to reach agreement concerning another farmer who had set about ploughing his marshes in the Yare valley. This drew the Prime Minister into the debate. Mrs Thatcher called in the respective ministers, and insisted that an exceptional Planning Direction to prevent drainage be imposed upon the farmer in question.[101]

The lonely church at Aughton in the heart of the Derwent Ings. Permission to drain the Ings was refused in an historic decision made in 1984.

A wetland newly created by the Severn–Trent Water Authority. River Arrow, Warwickshire.

In the wake of drainage on Halvergate came the Friends of the Earth. Photo: Ralph de Rijke.

That autumn, Patrick Jenkin, Secretary of State for the Environment, visited the Halvergate marshes to see for himself the landscape which was creating so much controversy and expense. The wetland behaved true to form. It was one of those November days when it never really gets light. The Government party arrived at the Berney Arms halt by special train, whence they were to be conveyed across the marsh by Land-Rover. On alighting from the train, they discovered the chartered Land-Rover up to its axles in slime, while a second one was wrapped around a stubborn gatepost, thus trapping the minister in the marshes. Fortunately, his impression appears to have been that Halvergate was one of the last untamed wildernesses in the country; and at a press conference in Norwich, he observed that ploughing the marshes would represent 'a major tragedy and a failure for conservation'.

With so much at stake, and in the public eye, something had to be seen to be done. In March 1985 the Broads Grazing Marsh Conservation Scheme was launched. This scheme, which may just possibly prove to be a revolutionary approach to farming the landscape, is funded jointly by the Countryside Commission and the Ministry of Agriculture, and offers all landowners a flat fee of £50 an acre (approximately one-third of what they would have obtained under the SSSI provisions of the Wildlife and Countryside Act) to retain low-gear livestock farming on the marshes. These arrangements have been fine-tuned by the designation of the Broads as an Environmentally Sensitive Area (ESA), sanctioned by the Common Market. In Spring 1987 nine ESAs, including both the Broads and the Somerset Levels, came into operation. If these prove a success, then future candidates, which include rivers and wetlands of importance, may be Anglesey, Constable country, and the Test valley in Hampshire. Ten per cent of the land within the ESA was affected by a second-tier payment of £80 an acre, which was brought in to encourage higher water-levels, less intensive grazing, and even less use of fertilizer, monitored by the Ministry of Agriculture through the remote surveillance techniques of false-colour photography.

By the end of 1986 a great many farmers had come into the original grazing marsh scheme. This means that at last the boot is on the other foot in the contest between graziers and arable farmers on the Halvergate marshes. If one farmer ploughs, he will prejudice the whole payment arrangement whereby the rest of the farming community benefits. Another factor working in favour of this scheme is the physical collapse of the soil structure on some drained and ploughed land, due to deflocculation on the clay land, together with acid sulphate problems on the peats. Above all, these payments, which are effectively a social, as well as a landscape, subsidy, are seen as part of a fair scheme, and equity has not exactly been at a premium in the Norfolk marshes in the 1980s.

None the less, although such schemes represent a major breakthrough, problems remain. Fine control of water-levels, especially under the terms of the Somerset ESA, do not appear to lie securely in the hands of the environmentalists, or even within the control of senior staff in the Ministry of Agriculture. In East Anglia, as in Somerset, a good future for the dairy industry is critically important for the survival of the grazed landscape. Unless all farmers within ESAs take up the scheme, then the patchwork of ploughland and pasture will mean that the schemes fail in landscape terms, and, even more important, total management of high water-levels to achieve the common purpose of grazing will be impossible. There is also the old question of cost. The Somerset Levels ESA promises to absorb £1.5 million of the annual national cake of £4.3 million for ESAs in England and Wales in 1987–8. Last, there is the fact that ESAs are small islands of integrated landscape management within the total framework of the rest of the countryside, where uncontrolled habitat removal continues. In this respect, they might be renamed PSAs—Politically Sensitive Areas—since they represent a sticking plaster to the unreliable system of voluntary agreement in the countryside.

## The Future for Land Drainage

In the winter of 1984/5 the Government announced major cuts in finance for land drainage, reducing the proposed expenditure for 1985–6 by £12 million, and announcing that urban and coastal flood relief should be given strict priority over drainage for agriculture, and that direct beneficiaries should pay more for the land-drainage benefits they receive.[102] In addition, the percentage levels of support from central government for internal drainage board and water authority schemes were reduced. It was commonly said among environmentalists that, as a real force to be reckoned with in the countryside, land drainage was dead. Such confidence could prove premature, however. Although many capital schemes have been cut, the issue of whether to call a halt to extensive river maintenance has not yet been addressed. Water authorities are also going ahead with work in some places without grant aid from the Ministry of Agriculture; and farmers can offset quite major drainage works against tax by calling it maintenance expenditure. One commentator has observed that the Ministry of Agriculture has managed to dodge its critics by moving 'forthrightly into coastal protection and sea defence, on the back of which much traditional drainage work could be funded'[103]—a spine-chilling thought, in light of the Government's announcement in August 1986 that it was granting an increase of £16.7 million for coastal defence.[104] The Chernobyl disaster in 1986 has vastly increased the momentum behind long-standing proposals for barrages on the Severn and the Mersey, as well as across Morecambe Bay and the Wash. With clean energy now balanced against the effects on estuarine ecology and birdlife, this presents a real dilemma for conservationists. Furthermore, if and when our food surplus is reduced and North Sea oil begins to run out, the clamour for more vigorous agricultural production could always begin again.

In February 1988 a major factor encouraging drainage was at last subjected to the first stages of reform. Common Market leaders accepted an annual ceiling for cereals, which were previously protected with an artificial price level however much farmers produced. In the mid-1980s, with milk quotas in existence and the threat, but no immediate certainty, of quotas on grain, farmers were naturally encouraged to plough every acre in order to make money while the going was good. As many farmers went out of business, some of the largest barley barons were tempted to buy up the land cheap to speculate immediately in grain, or else were waiting until economic circumstances offered renewed rewards for further drainage and intensive agriculture. It remains to be seen how the fine detail of the new agreement will work out in practice, but it does appear that the great corn boom, which has had such a disastrous effect on the English landscape, is finally over.

The political structure, by which land drainage is administered both within the water authorities and within the IDBs, and which has encouraged drainage, remains unaltered however. This, together with the financing and accountability of land drainage, was the subject of a long-awaited Government Green Paper which came out in March 1985.[105] This document, which was widely viewed by those within the water industry as saying very little, was described by the Council for the Protection of Rural England as 'tinkering with what appears to us to be an outmoded policy'.[106] The Green Paper is a consultation document, leading the way for legislation now delayed until after water privatization which will in turn affect it. In the mean time rates reform current in 1989 may also affect IDBs. When changes go through, the agricultural lobby will be anxious to ensure minimum reform. It is important, therefore, that the environmentalists press the Government into addressing the problems affecting our rivers and wetlands. The Royal Society for the Protection of Birds and the Council for the Protection of Rural England have each produced a useful analysis of the issue in their formal responses to the Government's Green Paper.[107]

The Green Paper maintains that 'a special structure for land drainage is still needed within the Water Authorities', but it raises the question of whether local land drainage committees (LLDCs) are necessary. These committees, frequently dominated by chairmen of the IDBs in their region, have been a vehicle for the vested interests of local large-scale farmers in promoting drainage schemes to be financed via the central funds of the Ministry of Agriculture and the water authorities. The Royal Society for the Protection of Birds has commented that 'LLDCs should be abolished. If a particular need for their retention can be demonstrated, however, then they should be non-statutory advisory committees.' The society also recommends that the regional land drainage committees become non-statutory and advisory, as are other key water authority committees. At last, then, what even the Anglian Water Authority, in its response to the Green Paper, has described as the land drainage 'State within a State' would lose its unique privileges. However, privatization now promises to change the ground rules whereby land drainage remains within the water authorities at all.

But perhaps the most disappointing evasion in the Government's document is its statement that there is 'a continuing role for IDBs'. The Green Paper has endorsed the need to reform the IDBs' most outrageous archaisms, such as voting strength proportional to land owned. These are an embarrassment to the drainage lobby, and the boards' umbrella group, the Association of Drainage Authorities, welcomes their removal as vigorously as do the conservationists. No one can dispute that local drainage expertise found in the boards will continue to be invaluable on advisory bodies, facilitating practical management of water-levels, essential for both wetland habitat and farming practice. What is wrong is for local agricultural interests to retain exclusive executive power over landscapes which are of national, and even international, importance, while also retaining access to national funds available from the Ministry of Agriculture. A far wiser solution, propounded by the Royal Society for the Protection of Birds and the Council for the Protection of Rural England, would be for water authorities to take over the most important drains currently administered by the boards, and for the smaller drains to become the responsiblity of individual landowners, just like the myriad other small ditches which are already maintained by the farmers who own them. In this way, IDBs could be abolished.

Objections to abolition have been twofold. A few environmentalists, notably those at the NCC, worry that a take-over of the boards by the water authorities would inject lavish, and so environmentally destructive, authority resources into the maintenance of drainage systems which were previously subject to inexpensive management by the boards. In fact, case histories of the Somerset Levels and Halvergate have shown how the specific remit of some IDBs has encouraged aggressive, high-gear drainage proposals. Meanwhile, with every passing year, the water authorities are proving themselves more responsible environmentally, and at the same time less free and easy with taxpayers' money in pursuit of drainage. The Association of Drainage Authorities has launched a laudable environmental campaign amongst its member boards entitled 'Find a Drain', whereby redundant ditch systems are set aside to be taken over by conservation organizations as aquatic habitat or else to be managed sympathetically by the boards themselves. None the less, the level of expenditure of IDBs to enhance the landscapes in their care is unlikely to match that available in the water authorities, which can truly fulfil the ideals of the enhancement clause of the Wildlife and Countryside Act. In the winter of 1986/7 the Severn–Trent Water Authority instigated the planting of 8 miles of willow trees, together with dyke widening for aquatic habitat, on the Hatfield Chase IDB which it administers. The IDB itself contributed £100 towards the water authority's input of £25,000. Yet this scheme could not be judged as a wild extravagance by the water authority. It was part of the 5 per cent annual allocation for landscape out of the authority's total land-drainage budget, which in turn amounts to an average of only 5 per cent of the authority's total annual expenditure. The benefits of this scheme have not only been welcomed by the local farming community; they can also be seen as the rebuilding of a fen landscape which is part of the national heritage of everyone. The Tavistock Report, upon which the Government's Green Paper relies heavily, admits that IDBs are 'not equipped in their organisational form, their modest resources, and their accountability structure' to reconcile major environmental and drainage conflicts.[108] With one clerk, frequently shared between many of them, it is unthinkable that IDBs could afford to employ the environmental staff which water authorities are increasingly taking on to enhance the special landscapes in their care.

The second major objection to the abolition of IDBs has been the expense of alternatives. However, the North-West Water Authority, which maintains the watercourses previously administered by eleven IDBs, and the Colchester division of the Anglian Water Authority, which also administers IDBs within its area, have not been crippled by the extra costs. If the IDB system of financing were to be fully integrated within a revised water authority system for financing agricultural land drainage, additional costs should not prove prohibitive. The ambitions of those on IDBs have been a major impetus behind the controversial drainage proposals in Somerset, Halvergate, and the Derwent Ings, which have consumed precious ministerial time; and they have also played a part in promoting such extravagant agricultural schemes as that still current in 1987 on the river Soar. To abolish the executive power of IDBs would be, in some cases, to save the nation from wasting its money on accelerated drainage, which adds to the grain mountain; and, at the same time, it would secure a better environment for many of our lowland rivers and wetlands.

The best security for these landscapes, however, lies in the final comments made by the Royal Society for the Protection of Birds and the Council for the Protection of Rural England on the Government's Green Paper: the need for proper planning controls over deep drainage, all major capital drainage schemes, and also heavy river maintenance. These operations have such a major impact upon the landscape that the total lack of effective control of them appears anomalous. Furthermore, because large-scale drainage is one of the most influential factors for change in the countryside, planning exercised on such operations, as it is in Germany, does not imply restrictions on all other farming activities, a prospect which has long alarmed the agricultural community.

*Privatization*

Following the general election in 1987, the Tory Government has set about privatizing the water industry. In February 1986 the Government published a White Paper announcing its proposals to privatize water,[109] followed in April by its package for the environment under privatization, issued by Minister for the Environment John Patten, and entitled 'The Water Environment. The Next Steps'.[110] With £7 billion at stake and major apprehension concerning privatization in the Press focused on environmental issues, the Government made much of the opportunity to strengthen environmental controls. 'The Next Steps' encouragingly advocated a detailed government code of practice, to ensure that existing best environmental practice in the water industry be adopted as normal practice. This opens up prospects for real consolidation of the work of the previous decade, which has slowly set standards for landscape enhancement on rivers and wetlands.

Worries remain that the code, which avoids 'unnecessary detail', may actually fall down on the necessary detail of money and staff needed to achieve the highly desirable standards it is now encouraging nation-wide. This is especially the case, given the high cost of pollution control. A river system with all the margins and trees one could want will not be very satisfactory if it is biologically dead.

In 1986, the Government shelved its plans for water privatization, but in 1987, just before the General Election, the Tories announced their proposal to create a new National Rivers Authority, to retain in the public sector the functions of water conservation and resource planning, pollution control, fisheries, land drainage and flood protection, and navigation. The utility functions of water supply, sewerage, and sewage treatment and disposal were to remain the responsibilities of the public limited companies, which would be transferred to the private sector.

Under this new plan some common ground is emerging between drainage men and environmentalists, who both look to a National Rivers Authority as the most practical way of achieving sound river management. Conservationists' anxieties that a cost-conscious privatized water industry would reduce environmental standards on the rivers may be partially allayed by the existence of a publicly funded body responsible for the river environment. Traditional land-drainage men, on the other hand, are inclined to see the NRA as a return at last to the old River Boards—a belated but none the less satisfactory victory in the battle they fought and lost in 1974 to achieve an independent drainage authority.

Whether either group will be disappointed depends upon the crucially important details upon which the Green Paper for the NRA remains notably vague.[111] There is a risk that such an authority could be nothing more than a small skeleton organization without effective power, relying on the privatized water authorities to do the real work on an agency basis. This would clearly be far less satisfactory than an authority with sufficient staff and funds to ensure not only that environmental standards are maintained where they already exist, but also that those standards are raised in areas where river enhancement is still rudimentary. Then at last a high standard of environmentally sympathetic river management could be achieved throughout the country.

The second risk is that the regional divisions of the NRA will be run by the present water authority land-drainage committees unchecked by the wider interests with which they have to compete in the existing water authorities. While it is reasonable that agriculture should always have a strong voice in drainage matters, it would be undesirable to allow a system to develop which might tempt farmers sitting both on the IDBs and the NRA committees to push through drainage which is in their own but nobody else's interest. The Green Paper, while suggesting that the Regional and Local Land Drainage Committees can be readily adapted for the NRA, accepts that membership of the committees should be drawn from a wider field than agriculture, now that benefits to farming have ceased to be the main priority for land drainage.

While the Government has accepted that the NRA should retain the same duty as the water authorities to further conservation, that duty together with the power to spend money on conservation only applies, according to the strict letter of the law, as part of the authorities' main function of land drainage; thus you can enhance a river only when you attack it. A river which might have been stripped out ten years before cannot at present benefit from tree planting unless there is a need to go in and dredge it again. This limited power is in contrast to the free-standing power to promote recreation conferred on water authorities by the 1973 Water Act. Water authorities can construct a car park beside a river any time they like but they have to wait until the machines are going through before they are empowered to create an otter holt. While the legislation is being drafted this anomaly should be corrected so that the NRA's powers to spend money on conservation match the existing powers for recreation.

The setting-up of a National Rivers Authority thus presents a golden opportunity to achieve proper environmental management of all our rivers and to put the administration of land drainage on a more democratic foundation. In such circumstances, the environmentalists' comments on the 1985 Land Drainage Green Paper can be applied no longer to the water authorities but instead to the National Rivers Authority, which could also take over the more important drains administered by IDBs. By late 1988 the Parliamentary debate will be under way to decide whether the ground rules which dictate enhancement and security for our rivers and wetlands will either be upgraded or simply whittled away.

### The River Blackwater

Such are the possibilities for sympathetic river management which appear on the horizon in the late 1980s. Meanwhile, if anyone needs convincing that the battle for a proper treatment of rivers and wetlands is still far from over in the United Kingdom, a visit to Northern Ireland is, as in so many other ways, a salutary experience. The river Blackwater, which flows into Lough Neagh, is one of Ireland's major rivers, forming over some reaches the border between the North and the South, and, together with its numerous tributaries, covering a catchment area of approximately 570 square miles.[112] In the 1960s the lower reaches around Lough Neagh were drained; and in 1982 a £32-million second stage was approved, with money provided by the EEC to promote cross-border harmony. In 1986 the scheme was well under way, affecting in Northern Ireland alone around 218 miles of watercourse, which was stripped of vegetation over many reaches.[113] The river engineering scheme actively promoted the drainage of over thirty-two wetlands which lie among the low humpy hills of the drumlin country of Tyrone and Armagh. Although £1 million is being set aside for tree planting beside the Blackwater, the sheer scale and suddenness of impact of the operation upon this exceptionally lovely landscape, combined with the lack of protection for many adjacent wetlands, make the Blackwater scheme something which would no longer be environmentally or politically acceptable across the water in England and Wales.

The Blackwater reflects all the tragic carelessness which large-scale land drainage represents at its worst. At the same time, all the issues and real human dilemmas which have underlain our generation's endeavours to tame the flood are emphatically underscored. In Northern Ireland, an environmental lobby, together with the necessary resources and legislation to protect the landscape, is either absent or quite inadequate to combat the massive momentum of Common Market-funded agriculture. In nature conservation, as in so much else, the English have failed the Irish people. In Northern Ireland, there is no Nature Conservancy Council; the enhancement clause of the 1981 Wildlife and Countryside Act does not apply; and the role of the water authorities and the Ministry of Agriculture, which in the rest of the United Kingdom have shown increasing environmental sympathy, is taken by the more autonomous Department of Agriculture of Northern Ireland. The Department of Environment of Northern Ireland rep-

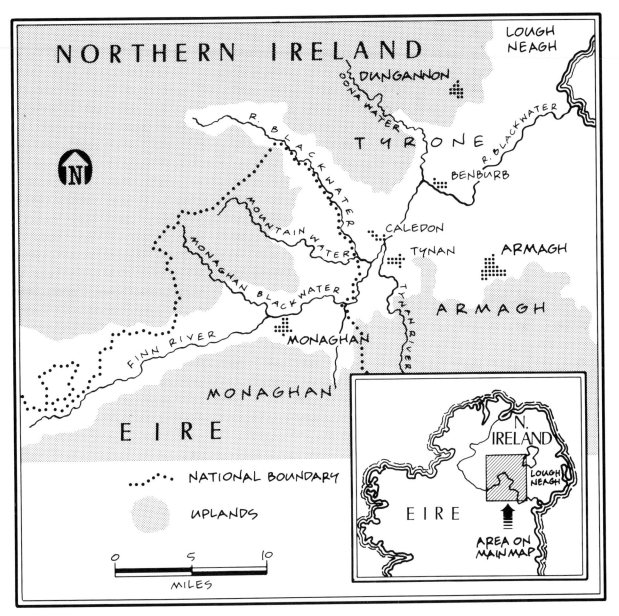

The River Blackwater and its tributaries.

resents statutory conservation interests in the province, and is currently re-designating Northern Ireland's equivalent of SSSIs, known before 1985 as ASIs, and now as ASSIs. It is indicative that in 1986 there are only 50 such protected areas in Northern Ireland, whereas there should be well over 200 if you compare the area of land to any similar region in England and Wales. Few wetlands are protected. Indeed, most protected sites in Ireland are such places as sea-bird cliffs, which scarcely pose a threat to progressive agriculture or any other developments.

General acceptance in the rest of the United Kingdom that agricultural land-drainage schemes should cease to be regarded as a priority has by no means arrived in Northern Ireland, where the system devised by the Department of Agriculture for assessing the costs of agricultural drainage, together with the secrecy with which it is shrouded, has been heavily criticized by the Northern Ireland Economic Council.[114] The harvests of drainage in this wet climate will be more milk and butter to add to the European surplus. Yet in 1987, investment promises to continue unabated. The Royal Society for the Protection of Birds has said of these proposals that 'their environmental impact would be severe; the economic benefit negligible'.[115]

The river Blackwater is also a classic example of the physical collapse of a river system as a result of large-scale engineering.[116] The unstable gravels, sands, and silts of the Blackwater valley have caused the newly engineered river banks to slip, creating massive cavities, and causing stiles and fences to slide into the brink. This has necessitated the return of further armies of engineers to patch up the increasingly uncontrollable watercourse with cement and stone—something scarcely allowed for in the initial budgets. In addition, greatly increased sediment washed downstream from the area covered by the scheme is now creating the need for further maintenance operations on the lower reaches of the Blackwater.

In Ireland, where agriculture lies at the heart of the economy, drainage has been described as a matter of 'national pride', the value of which it would be 'unpatriotic' to question, and one commentator has written of drainage there: 'It has been a feature of public works programmes for nearly a century and a half and there is almost an element of heresy in questioning its value. Yet the lack of rigorous economic analysis invites precisely such criticism.'[117] In the Republic, the EEC has targeted money on to the western part, where uncontrolled drainage continues to reduce the quality of fishing and the tourist potential of some areas, as well as destroying outstanding wetland features such as the shallow lakes, known as turloughs, and endangering internationally important populations of Greenland white-fronted geese.

The short-term benefits for drainage contractors and local landowners from a scheme such as the Blackwater, where, in the floods of 1985, there were stories of farmers up to their waists in the river all night, rescuing their cattle, are undeniable. But in Ireland there are few people to question the long-term benefits of such a scheme. The old problem of outsiders telling local farmers what is best for them is especially acute in the vexed political circumstances of the Blackwater valley. Such is the level of violence in the province that conservationists risk rather more than effigy burning, and the English, who have so long insulted the people of Ireland with the epithet 'bog-trotter', are not best placed to rush in and tell them to stop draining their bogs. If the wetlands and rivers of Ireland are to be saved, they must clearly be saved by the Irish for the benefit of Ireland.

The valley of the Blackwater is the more hell for looking like heaven. From the spectacular castle-crowned gorge at Benburb, you can travel along the wide

River bank before and after engineering on the Blackwater scheme. Over 200 miles of river are being affected.

Dipper.

foaming river up to Caledon, most musically named of villages, whose church spire of silvery stone commands a little hill above a reedy mere. Grey-green lichen frosts the elegant stone bridges and the great lime trees. Continuously, in your ears, like an ever-beating sea, sounds the clamorous cawing of the rooks. In this spectacular landscape, the river, which is the trench between two warring factions, is being physically deepened by an agricultural and political machinery which is also out of control. In Caledon the pavements are picked out in red, white, and blue, and little children are led by the hand up the village street, where foot patrols on exercise run for cover. On the walls is painted the red hand of Ulster, the grim symbol which recalls how an Irish chief cut off his own hand and threw it ahead of him to reach a boundary ahead of his rival. Perhaps that boundary was a river.

All along the Blackwater and its tributaries, dredgers and cranes clear bankside vegetation which once provided shelter for otters, dippers, and grey wagtails. Ash trees, hazel, and a delicate cascade of snowdrops are dismembered with equal attention, and the limbs and trunks of trees are piled high, to rot beside the river. With the single exception of the goose grounds at Annaghroe, the riverside wetlands, which once supported snipe, water-rail, mallard, and meadow pipit, together with such rarities as marsh fern and fen bedstraw, are left to drain away without having received even the benefit of a thorough survey, let alone formal protection.[118] When the main engineering contract ends, local contractors will move in to the remaining tributary streams; and as the banks on the engineered rivers cave in, more concrete will be brought in to patch up the wounds. It is clear, however, that mismanagement of our rivers is only part of a far greater failure. At Tynan, in 1986, written on a wall beside the Blackwater is the date 1690, the year of the Battle of the Boyne. Near by, the engineers' portacabin is dwarfed among the tangled wreckage of torn-up trees, and in the woods behind are the charred remains of Tynan Abbey, where a few years ago Sir Norman Strong and his son were shot, and the house was burned to the ground. Looking across the palpable ruin of the engineering tackle, you know that the burned house lies somewhere beyond the gleaming silver lakes, the haunt of heron and wild duck, which stretch away towards the border and which are about to be unplugged.

## When the Next Flood Comes

The river Blackwater represents the real nightmare of what could happen to our rivers and wetlands in England and Wales if an unbridled agricultural lobby ever again manages to push through a national policy of drainage at all costs in the aftermath of floods and food shortages. As it is, the face of river management has changed beyond recognition in the decade since the water authorities were set up in 1974. On the issue of wetland drainage, we have seen conservationists gain strength with every passing year; and just as in the nineteenth century the hills were for the first time appreciated as magnificent landscapes, rather than wastes, our generation has rediscovered the levels and wetlands of the lowlands as another special kind of wilderness, with a value not only for wildlife but also for the human spirit. As conservationists buy pasture in such places as the north Kent marshes and Halvergate, they are starting to actively manage wetlands to raise the water over the land. Both conservationists and farmers share a common interest in the wise management of water, and they must both recognize that there will always be a need to tame the flood.

The way in which water forever waits for its moment to flood is well illustrated by a common plant of damp places called the amphibious bistort. Under normal conditions of rough pasture, it is an ugly duckling of a plant, its scorched ragged leaves scarcely noticeable among the grasses. But if you dig a pond where it occurs, it will miraculously extend out into the water, changing shape as it does so, and develop specialized aquatic leaves and rosy flowers, which neatly splay out over the surface in patterns of almost oriental elegance. When the real floods come, the amphibious bistort comes into its own. A photograph of the East Coast disaster of 1953 shows not only the abandoned implements of agriculture marooned in the glittering fields of water, but also the exuberant foliage of the bistort, which is always waiting for another flood.

As we manage our watercourses in an effort to minimize such disasters, we owe it to our children to rebuild the rivers, so many of which have been devastated. Engineers have now begun to take on a proper brief, in which the environment and the necessary management to control flooding are properly integrated. In doing so, they may be said to be refining the 'palimpsest' of the river valleys. A palimpsest, my dictionary tells me, is 'a manuscript, the original writing on which has been erased to make room for a second'. In other words, writing on writing, but with, often as not, some of the original writing showing through. An address book can be a good palimpsest. As people move house or

The flood may always return. Amphibious bistort flourishes in the place of agriculture, 1953. Photo: HMSO.

die, you score them out and write new names and addresses over the top. Thus, over the years, the book becomes a many-layered record of human contact. It is the same with the English landscape, quirky and profoundly humanized by the undesigned evolution of layer upon layer of contributions from every era since the Bronze Age. One contribution which we can make to that landscape is to return wildlife and elegance to our rivers.

'Improvement makes straight roads,' wrote William Blake, 'but the crooked roads,' he added, 'are roads of genius.'[119] How much more so, then, are rivers places of genius, in that they abhor the straight and narrow. Genius, surely, is something which cannot be reduced to a formula; and the genius of the English landscape is that it is as impossible to recreate it according to any theory of aesthetic composition as it would be idle to attempt to design a human face— or to force a river to obey the mathematics of Manning forever. Just as the task of river managers is to guide their streams on the right course, taking into account all the ecological and cultural aspects of their character as they do so, the challenge facing all those who manage our countryside is to evolve the palimpsest. That palimpsest is, and must continue to be, the product of active practical management; and nothing epitomizes the need for management in the countryside more than the flood, which, if not tamed partially, will rise to threaten us all.

# NOTES

## Chapter 1

1. *The Times*, 27 Oct. 1984.
2. W. Wordsworth, 'The River Duddon', in *Wordsworth: Poetical Works,* ed. T. Hutchinson and E. de Selincourt (Oxford University Press, 1936).
3. P. S. Corbet, C. Longfield and N. W. Moore, *Dragonflies* (Collins, 1960). See also H. B. N. Hynes, *The Ecology of Running Waters* (Liverpool University Press, 1970).
4. W. Shakespeare, Sonnet 15.
5. F. B. O'Connor *et al.*, *Otters 1977* (Nature Conservancy Council and Society for the Promotion of Nature Conservation, 1977). See also P. R. F. Chanin and D. J. Jefferies, 'The Decline of the Otter, *Lutra lutra, L.*, in Britain: An Analysis of Hunting Records and Discussion of Causes', *Biological Journal of the Linnean Society,* 10 (1978); C. F. Mason and S. M. Macdonald, *Otters: Ecology and Conservation* (Cambridge University Press, 1986).
6. E. Thomas, 'The Mill Water', in *The Collected Poems of Edward Thomas,* ed. R. G. Thomas (Oxford University Press, 1981).
7. Ibid.
8. R. Blythe, *Akenfield* (Penguin, 1969).
9. A. Brian, 'The Effect of Man-made Structures on the Distribution of Plants Growing in the River Lugg', *Transactions of the Woolhope Naturalists' Field Club,* 44, part 2 (1983).
10. E. Pollard, M. D. Hooper and N. W. Moore, *Hedges* (Collins, 1974).
11. G. Sturt, *The Wheelwright's Shop* (Cambridge University Press, 1923).
12. E. Carter, Farming and Wildlife Advisory Group Newsletter, 1982.
13. See Pollard *et al.*, above, n. 10.
14. K. W. Smith, 'Breeding Waders of Wet Meadows Survey 1982 and 1983', *Bird Study,* 30 (1983). See also O. Mitchell, 'The Breeding Status and Distribution of Snipe, Redshank and Yellow Wagtail in Sussex', *Sussex Bird Report,* 34 (1981).
15. D. M. Stoddard, 'Individual Range, Dispersion and Dispersal in a Population of Water Voles, *Arvicola terrestris L*', *Journal of Animal Ecology,* 39 (1970).
16. F. Pryor, 'Flag Fen', *Current Archaeology,* 96 (Apr. 1985).
17. D. A. Ratcliffe, *Objectives and Strategy for Nature Conservation in Great Britain* (Nature Conservancy Council, 1983).

## Chapter 2

1. C. Kingsley, *Prose Idylls* (1873).
2. E. Ekwall, *Concise Oxford Dictionary of English Place-Names* (Oxford University Press, 1951).
3. W. Dugdale, *History of Imbanking and Drayning of Diverse Fennes and Marshes* (1662).
4. Felix of Crowland, 'Vita Guthlaci' trans. W. de Gray Birch, in *Memorials of St. Guthlac of Crowland* (1881).
5. W. Lambarde, *A Perambulation of Kent* (1596).
6. H. C., *A Discourse Concerning the Drayning of Fennes and Surrounded Grounds in the Sixe Counteys of Norfolke, Suffolke, Cambridge, with the Isle of Ely, Huntingdon, Northampton and Lincolne* (1629).
7. S. Pepys, *Diary,* entry for 17 Sept. 1663.
8. C. Fiennes, *The Journeys of Celia Fiennes, 1685–1703* (Futura, 1983).
9. W. Hall, *A Chain of Incidents Relating to the State of the Fens from Earliest Accounts to the Present Time* (King's Lynn, 1812).
10. D. Defoe, *A Tour through England and Wales* (1722; repr. Everyman's Library, 1928).
11. G. Plaxton, 'Some Natural Observations in the parishes of Kinnardsley and Donnington in Shropshire', *Philosophical Transactions,* 25 (1673).
12. Defoe, above, n. 10.
13. A. A. Mills, 'Will-o'-the-Wisp', *Chemistry in Britain,* 16 (1980).
14. W. Watson, *An Historical Account of the Ancient Town and Port of Wisbech* (1827).

15. R. B. Grantham, *Report on the Floods in Somersetshire in 1872–1873* (1873).

16. W. McArthur, 'A Brief Story of English Malaria', *Transactions of the Royal Society of Tropical Medicine and Hygiene*, 46 (1952). Other species of mosquito have been suggested as carriers. See O. Rackham, *The History of the Countryside* (Dent, 1986).

17. F. Engels, *The Condition of the Working Class in England, 1845* (Panther, 1969). See also N. J. Barton, *The Lost Rivers of London, 1962* (Historical Publications Ltd., 1982).

18. G. Orwell, *The Road to Wigan Pier* (1937; repr. Penguin, 1962).

19. Fiennes, above, n. 8; my emphasis.

20. W. Elstobb, *An Historical Account of the Great Level of the Fens* (1793).

21. C. Vancouver, *General View of the Agriculture of the County of Cambridgeshire* (1794).

22. *The Anti-Projector; or the History of the Fen Project* (c.1646).

23. J. Thirsk, 'The Isle of Axholme before Vermuyden', *Agricultural Historical Review*, 1 (1953).

24. W. Cobbett, *Rural Rides* (1830; repr. Penguin, 1967).

25. J. R. Ravensdale, *Liable to Floods* (Cambridge University Press, 1974).

26. W. M. Palmer 'On the Cambridgeshire assize rolls', *Proc. Camb. Antiq. Soc.*, 3 (1896).

27. W. Camden, *Britannia* (1586; rev. 1637). Camden's spelling is 'puittes', so he would have called them 'pewits', not 'peewits'.

28. M. Drayton, *The Second Part, or a Continuance of Polyolbion* (1622).

29. M. Williams, *The Draining of the Somerset Levels* (Cambridge University Press, 1970).

30. Ibid.

31. W. H. Wheeler, *A History of the Fens of South Lincolnshire* (1794).

32. J. A. Sheppard, 'The Draining of the Marshlands of South Holderness and the Vale of York' (East Yorkshire Local History Society).

33. T. Stone, *Agriculture of the County of Lincoln* (1794).

34. B. Storer, *Sedgemoor: Its History and Natural History* (David and Charles, 1972).

35. T. Mackay, ed., *The Reminiscences of Albert Pell Sometime MP for South Leicestershire* (1908).

36. T. Fuller, *History of the University of Cambridge* (1655).

37. J. Bentham, *The Claim of Taxing the Navigations and Free Lands for the Drainage and Preservation of the Fens Considered* (1778).

38. Camden, above, n. 27.

39. Hammond, *Short Survey of Western Counties* (1635).

40. A. Young, *Agriculture of the County of Lincoln* (1813).

41. R. Gough, *The History of Myddle* (1700; repr. Penguin, 1981).

42. *Calendar of State Papers, Domestic, Chas. I,* ccxxx, 50 (1632); H. C. Darby, *The Changing Fenland* (Cambridge University Press, 1983).

43. R. T. Rowley, *The Shropshire Landscape* (Hodder and Stoughton, 1972).

44. G. Stovin, *The History of the Drainage of the Great Levil of Hatfield Chase in the Counties of York, Lincoln, and Nottingham* (1761).

45. In *Monumenta Historica Britannica*, ed. H. Petrie and J. Sharpe (1848).

46. Wheeler, above, n. 31.

47. A. Motion, 'Inland', 1975 Newdigate Prize Poem, in *The Pleasure Steamers* (Carcanet, 1978).

48. J. Billingsley, *A General View of the Agriculture of the County of Somerset* (1794).

49. J. Thirsk, 'Fenland Farming in the Sixteenth Century', University of Leicester, Department of English Local History; occasional papers, ser. 3 (1953).

50. C. J. Bond, 'Otmoor', in *The Evolution of Marshland Landscapes* (Oxford Univ. Dept. for External Studies, 1981).

51. Billingsley, above, n. 48. See also idem, 'On the Uselessness of Commons to the Poor', *Annals of Agriculture,* 31 (1798); idem, 'On the Best Method of Inclosing, Dividing and Cultivating Waste Lands', *Letters and Papers of the Bath and West of England Society,* 11, (1809).

52. Quoted in R. Millward, *Lancashire* (Hodder and Stoughton, 1955).

53. Young, above, n. 40; my emphasis.

## Chapter 3

1. Quoted by H. G. Richardson, 'The Early History of Commissions of Sewers', *English Historical Review*, 34 (1919).

2. T. W. Potter, 'Marshland and Drainage in the Classical World', in *The Evolution of Marshland Landscapes* (Oxford Univ. Dept. for External Studies, 1981).

3. P. Salway *et al., The Fenland in Roman Times* (The Royal Geographical Research Series, 5, 1970).

4. J. Harrison and P. Grant, *The Thames Transformed* (Andre Deutsch, 1976).

5. C. J. Bond, 'The Marshlands of Malvern Chase', in *The Evolution of Marshland Landscapes* (Oxford Univ. Dept. for External Studies, 1981).

6. D. Hall, 'The Changing Landscape of the Cambridgeshire Silt Fens', *Landscape History*, 3 (1982).

7. H. Neilson, *The Cartulary and Terrier of the Priory of Bilsington, Kent* (Oxford University Press, for the British Academy, 1928). See also N. P. Brooks, 'Romney Marsh in the Early Middle Ages', in *The Evolution of Marshland Landscapes* (Oxford Univ. Dept. for External Studies, 1981).

8. Bond, 'Otmoor', above, ch. 2, n. 50.

9. J. M. Steane, *The Northamptonshire Landscape* (Hodder and Stoughton, 1974).

10. Williams, above, ch. 2, n. 29.

11. J. Gardner, *The Life and Times and Chaucer* (Paladin, 1979).

12. P. Brandon, *The Sussex Landscape* (Hodder and Stoughton, 1974).

13. P. R. Edwards, 'Drainage Operations in the Wealdmoors, Shropshire', in *The Evolution of Marshland Landscapes* (Oxford Univ. Dept. for External Studies, 1981).

14. A. L. Rowse, *The England of Elizabeth* (Macmillan, 1953).

15. Anon., *A True Report of Certaine Wonderfull Over-flowings of Waters, now lately in Summersetshire, Norfolke, and other Places of England* (1607).

16. Dugdale, above, ch. 2, n. 3.

17. *Calendar of State Papers, Domestic, Jas. I*, xix, 47 (1606).

18. L. E. Harris, *Vermuyden and the Fens* (Cleaver-Hume, 1953).

19. Stovin, above, ch. 2, n. 44, quoting, '*Attorn'd Regis. Vermuyden vs Torksey et al.*'

20. J. Korthals-Altes, *Sir Cornelius Vermuyden* (Williams and Norgate, 1925).

21. R. Locke, 'A Historical Account of the Marshlands of the County of Somerset', *Letters and Papers, Bath and West of England Society*, (1796). See also M. Williams, *The Draining of the Somerset Levels* (Cambridge University Press, 1970).

22. See Dugdale, above, ch. 2, n. 3.

23. Wheeler, above, ch. 2, n. 31. For the best accounts of the drainage and development of the Fens, see H. C. Darby, *The Changing Fenland* (Cambridge University Press, 1983), and H. Godwin, *Fenland: Its Ancient Past and Uncertain Future* (Cambridge University Press, 1978).

24. *Calendar of State Papers, Domestic, Chas. I*, cccxcii, 28 (1638).

25. Dugdale, above, ch. 2, n. 3; *Calendar of State Papers, Domestic, 1631–33*. See also C. Hill, *God's Englishman, Oliver Cromwell and the English Revolution* (Weidenfeld and Nicolson, 1970, Pelican, 1973).

26. See Harris, above, n. 18.

27. *Calendar of State Papers, Domestic, Commonwealth*, xxix, 1653.

28. S. Fortrey, attrib., *History or Narrative of the Great Level of the Fens called Bedford Level* (1685).

29. See Harris, above, n. 18.

30. T. Randolph, *The Muse's Looking Glass* (1643).

31. S. Marshall, *Fenland Chronicle* (Cambridge University Press, 1967).

32. *Calendar of State Papers, Domestic, Chas. II*, cccxxxii, 3 (1673).

33. C. Taylor, *Dorset* (Hodder and Stoughton, 1970).

34. P. Bateson, *An Answer to some Objections by Hatton Berners, Esq.* (1710).

35. See Millward, above, ch. 2, n. 52. Also C. Leigh, *The Natural History of Lancashire, Cheshire and the Peak in Derbyshire* (1700).

36. B. Trafford, 'Field Drainage', *Journal of the Royal Agricultural Society of England*, 131 (1970).

37. J. Johnstone, *An Account of the Most Approved Mode of Draining Land According to the System Practised by Mr Joseph Elkington, Late of Princethorp in the County of Warwick* (1797).

38. *Farmers' Magazine*, Feb. 1807.

39. J. Goodier, 'Chat Moss. Its Reclamation, its Pioneers', in *Lectures 1970–71* (Eccles and District History Society, 1971).

40. E. Beazley, *Madocks and the Wonder of Wales* (Faber, 1967; P. and Q., 1985).

41. T. L. Peacock, *Headlong Hall* (1816; repr. Everyman, 1966).

42. E. Robinson, ed., *Selected Poems and Prose of John Clare* (Oxford University Press, 1967).

43. *Journals of the House of Commons*, xxxiii (1775).

44. Bond, 'Otmoor,' see above, ch. 2, n. 50. See also H. G. Hobson and K. L. H. Price, 'Otmoor and its Seven Towns' (booklet printed by A. T. Broome and Son, 1961); R. Mabey, *In a Green Shade* (Hutchinson, 1983).

45. A. G. Ruston, 'Land Improvement by Warping', *Journal of the Ministry of Agriculture*, 41 (1934).

46. J. A. Clarke, *Fen Sketches* (1852). See also N. Harvey, *The Industrial Archaeology of Farming in England and Wales* (Batsford, 1980); idem, 'Ditches, Dykes and Deep Drainage', Young Farmers' Club Booklet, 29 (1956).

47. H. White, 'A Detailed Report of the Drainage by Steam-power of a Portion of Martin Mere,

Lancashire', *Journal of the Royal Agricultural Society*, 31 (1853).

48. W. T. Bree, 'Recollections of a Morning's Ramble in the Whittlesea Fens', *Phytologist*, 4 (1851). See also S. H. Miller and S. B. J. Skertchly, *The Fenland* (1878).

49. Wheldrake Ings Account Book, 1867–1935, unpublished manuscript.

50. In *The Richmond Papers*, ed. E. M. W. Stirling (1926); quoted by P. Henderson, *William Morris* (Thames and Hudson, 1967).

51. C. H. J. Clayton, 'Land Drainage Works for the Relief of Unemployment', *Journal of Agriculture*, 29 (1922).

52. H. V. Garner, 'The "Buckeye" Ditcher for Land Drainage: Trial in Cambridgeshire', *Journal of Ministry of Agriculture*, 28 (1921).

53. S. Webb, *English Local Government* (Longmans Green, 1922).

54. B. A. Keen, 'Land Drainage: The Area of Benefit', *Journal of Agriculture*, 43 (1936). See also A. Dobson and H. Hull, *The Land Drainage Act, 1930* (Oxford University Press, 1931).

55. E. A. G. Johnson, 'Land Drainage in England and Wales', *Proceedings of the Institution of Civil Engineers*, 3 (Dec. 1954).

56. J. Wentworth-Day, *A History of the Fens* (Harrap, 1954). See also H. J. Mason, *Harvest Home. The Story of the Great Floods of 1947* (Providence Press, Ely, 1985).

57. E. R. Delderfield, *The Lynmouth Flood Disaster* (Raleigh Press, 1953).

58. D. Summers, *The East Coast Floods* (David and Charles, 1978).

59. E. A. R. Ennion, *Adventurers Fen* (Methuen, 1942).

60. A. H. V. Bloom, *The Farm in the Fen* (Faber and Faber, 1944).

## Chapter 4

1. B. D. Trafford, 'Recent Progress in Field Drainage: Part I', *Journal of the Royal Agricultural Society*, 138 (1977); also A. Blenkharn, 'Benefits and demand of land drainage', *Water*, Nov. 1978.

2. D. Baldock, *Wetland Drainage in Europe, The Effects of Agricultural Policy in Four EEC Countries* (International Institute for Environment and Development and Institute for European Environmental Policy, 1984); J. K. Bowers, 'Cost-benefit analysis of wetland drainage', *Environment and Planning A*, 15 (1983); E. C. Penning-Rowsell *et al.*, *Floods and Drainage* (Allen and Unwin, 1986); B. D. Trafford, *The Background to Land Drainage Policies and Practices in England and Wales* (Ecole Polytechnique, Paris, 1982). There is a lack of absolute consensus on precise figures among the authorities.

3. *The Times*, 28 Sept. 1985.

4. J. F. Bird, 'Geomorphological Implications of Flood Control Measures: Lang Lang River, Victoria', *Australian Geographical Studies*, 18 (1980).

5. J. Lewin, 'Initiation of Bedforms and Meanders in Coarse-grained Sediment', *Bulletin of the Geological Society of America*, 87 (1976).

6. G. J. Leeks, J. Lewin and M. D. Newson, 'Channel Change, Fluvial Geomorphology and River Engineering: The Case of the Afon Trannon, Mid-Wales', *Earth Surface Processes and Landforms*, 1986.

7. R. D. Hey, 'River Mechanics', *Journal of IWES*, 40 (1986).

8. A. Krause, 'On the Effect of Marginal Tree Rows with Respect to the Management of Small Lowland Streams', *Aquatic Botany*, 3 (1977).

9. S. S. D. Foster *et al.*, *The Groundwater Nitrate Problem* (British Geological Survey, 86, 1986).

10. R. P. C. Morgan, ed., *Soil Conservation: Problems and Prospects* (John Wiley, 1981).

11. R. Evans, 'Soil Erosion—The Disappearing Trick', in *Proceedings of National Agricultural Conference*, 1985; R. D. Hodges and C. Arden-Clarke, *Soil Erosion in Britain* (The Soil Association, 1986).

12. See A. A. Thorburn and B. D. Trafford, 'Iron Ochre in Drains', MAFF Land Drainage Service, Technical Bulletin, 76.1; Environmental Resource Management Limited *et al.*, Acid Sulphate Soils in Broadload (Broads Authority, BA, RS, 3); E. Long, 'Sour Taste in the East', *Farmers' Weekly*, 9 Nov. 1984.

13. The Ministry of Agriculture operates a classification for all farmland, in which grade 1 is the very best, and grade 5 is the most worthless for agriculture.

14. E. Turner, 'Saline and Acid Sulphate Soils, in *Proceedings of ADAS Wildlife and Countryside Conservation Conference, Nov. 1985* (ADAS, 1986); J. Hazelden, P. J. Loveland and R. G. Sturdy, 'Saline Soils in North Kent', *Soil Survey*, 14 (1986).

15. H. Hope, 'Drainage Challenge on Isle of Grain Estate,' *Farmers' Weekly*, 2 May 1986.

16. 'Ryde's Scale', *Estates Gazette*, Nov. 1984; 'Revised Annex to Ryde's Scale', *Estates Gazette*, Aug. 1985.

17. N. T. H. Holmes, *Wildlife Surveys of Rivers in Relation to River Management* (Water Research Centre, 1986).

18. E. A. G. Johnson, 'Land Drainage in England and Wales', *Proceedings of the Institution of Civil Engineers*, 3 (1954).

19. See J. K. Bowers and P. Cheshire, *Agriculture, the Countryside and Land Use, An Economic Critique* (Methuen, 1983); G. Williams, Land Drainage in England and Wales: An Interim Report (Royal Society for the Protection of Birds, 1983); A. Warren and F. B. Goldsmith, eds., *Conservation in Perspective* (John Wiley, 1981); J. K. Bowers, 'Cost-Benefit Analysis of Wetland Drainage', *Environment and Planning A*, 15 (1983).

20. Investment Appraisal of Arterial Drainage, Flood Protection and Sea Defence Schemes, Guidance for Drainage Authorities (MAFF Land and Water Service, 1985).

## Chapter 5

1. T. A. Rowell, 'History and Management of Wicken Fen' (University of Cambridge, unpublished thesis, 1983).

2. P. M. Wade, 'The Effect of Mechanical Excavators on the Drainage Channel Habitat', in *Proceedings of EWRS, 5th Symposium on Aquatic Weeds* (EWRS, 1978).

3. B. Storer, *Sedgemoor: Its History and Natural History* (David and Charles, 1972).

4. P. M. Wade and R. W. Edwards, 'The Effect of Channel Maintenance on the Aquatic Macrophytes of the Drainage Channels of the Monmouthshire Levels, South Wales, 1840–1976', *Aquatic Botany*, 8 (1980).

5. D. A. Scott, ed., *Managing Wetlands and Their Birds, A Manual of Wetland and Waterfowl Management* (International Waterfowl Research Bureau, Slimbridge, 1982).

6. G. J. Thomas, D. A. Allen and M. P. B. Grose, 'The Demography and Flora of the Ouse Washes, England', *Biological Conservation*, 21 (1981).

7. J. M. Coles and B. J. Orme, *Prehistory of the Somerset Levels* (Somerset Levels Project, 1980).

8. S. B. Edgington, 'Buckden Churchwardens' Accounts 1627–1774', *Records of Huntingdonshire*, 12 (1981).

9. J. Clare, 'The Moorhen's Nest', in *Selected Poems and Prose of John Clare*, ed. E. Robinson and G. Summerfield (Oxford University Press, 1967).

10. J. C. Willis and I. H. Burkill, 'Observations on the Flora of the Pollard Willows near Cambridge', *Proceedings of the Cambridge Philosophical Society*, 8 (1893).

11. O. Rackham, *Trees and Woodland in the British Landscape* (Dent, 1976).

12. C. K. Catchpole, 'Aspects of Behavioural Ecology in Two *Acrocephalus* Species' (Unpublished University of Nottingham thesis, 1970); R. J. Hornby, 'Aspects of Population Ecology in the Reed Bunting' (Unpublished University of Nottingham thesis, 1971).

13. K. G. Stott, 'Cultivation and Uses of Basket Willows', *Quarterly Journal of Forestry*, (Apr. 1956); idem, 'A Review of the British Basket-Willow Growing Industry' (Long Ashton Research Station, 1959).

14. H. E. Fitzrandolph and M. D. Hay, *The Rural Industries of England and Wales*, vol. 2, *Osier Growing and Basketry and Some Rural Factories* (Oxford University Press, 1926; repr. E. P. Publishing, 1977). For a good account of the Somerset osier industry, see also A. Nicolson, *Wetland. Life in the Somerset Levels* (Michael Joseph, 1986). Research and advice on willow cultivation in the United Kingdom has centred on Long Ashton Research Station, near Bristol, since 1922.

15. R. Boston and A. Heseltine, 'David Drew: Baskets' (Crafts Council, 1986).

16. Anon., 'The Great Level of the Fenns' (*c.*1680).

17. P. Downing *et al.*, 'Small Woods on Farms, A Report to the Countryside Commission by Dartington Amenity Research Trust', Countryside Commission Publication, 143, (1983). See also *Proceedings of Farming and Forestry Conference 1986* (Loughborough University of Technology, 1986).

18. See the following (papers read at European Seminar, Research and Development on Pulp, Paper and Board, Brussels, 19–21 Nov. 1986): I. F. Hendry, 'Wood as a Renewable Raw Material'; C. de Choudens, 'New Technology for Bleached High Yield Pulp Process'; A. Bosia, 'CTMP from Hardwoods and Their Possible Use'. See also these papers read at the EUCEPA conference, Florence, 1986: I. F. Hendry, 'Research Needs for the EEC Pulp and Paper Industry'; A. Bosia, D. Nisi and P. Cappalletto, 'CTMP from Hardwoods'.

19. G. H. McElroy and W. M. Dawson, 'Biomass from Short-Rotation Coppice Willow on Marginal Land' (Horticultural Centre, Loughgall, 1986).

20. A. H. V. Bloom, *The Farm in the Fen* (Faber and Faber, 1944).

21. J. Gerard, *The Herbal or Generall Historie of Plantes* (1597).

22. A. Huxley, *Green Inheritance* (Collins, 1984).

23. E. Launert, *The Hamlyn Guide to Edible and*

*Medicinal Plants of Britain and Northern Europe* (Hamlyn, 1981).

24. M. Rothschild and T. Clay, *Fleas, Flukes and Cuckoos* (Collins, 1952). See also B. Payton, 'History of Medicinal Leeching and Early Medical References', in *Neurobiology of the Leech,* ed. T. E. S. Müller, J. G. Nicholls and G. S. Stent (Cold Spring Harbour, 1981).

25. J. M. Elliott and P. A. Tullett, 'The Status of Medicinal Leech, *Hirudo medicinalis,* in Europe and Especially in the British Isles', *Biological Conservation,* 29 (1984); P. J. Wilkin, 'A Study of the Medicinal Leech, *Hirudo medicinalis,* with a Strategy for its Conservation (unpublished University of London thesis, 1987).

26. H. Debout and M. Provost., 'Le Marais de la Sangsurière', *Courrier de la Nature,* 74 (1981).

27. J. Elkington, 'How the Sucker Helped the Clot', *Guardian,* 'Futures', 28 Feb. 1985.

28. E. Maltby, *Waterlogged Wealth* (Earthscan, 1986); IUCN Conservation Monitoring Centre, *Directory of Wetlands of International Importance* (International Union for Conservation of Nature and Natural Resources, May 1984).

29. F. Pannier, 'Mangroves Impacted by Human-Induced Disturbances: A Case Study of the Orinoco Delta Mangrove Ecosystem', *Environmental Management,* 3 (1979).

30. R. E. Turner, R. G. Wetzel and D. F. Whigham, eds., *Wetlands, Ecology and Management* (National Institute of Ecology and International Science Publishers, Lucknow, India, 1980).

31. C. J. Bibby, 'Wintering Bitterns in Britain', *British Birds,* 74 (1981); J. C. U. Day and J. Wilson, 'Breeding Bitterns in Britain', *British Birds,* 71 (1978); idem, 'Status of Bitterns in Europe since 1976', *British Birds,* 74 (1981).

32. C. J. Bibby, 'Food Supply and Diet of the Bearded Tit', *Bird Study,* 28 (1981).

33. C. J. Bibby and J. Lunn, 'Conservation of Reed Beds and their Avifauna in England and Wales', *Biological Conservation,* 23 (1982).

34. C. Cator, Outline of the British Reed Industry (pers. comm. from the British Reed Growers Association, May 1986). See also S. M. Haslam and D. S. A. McDougall, *The Reed* (Norfolk Reed Growers Association, 1969); K. Darby, 'Thatch as a Modern Building Material', *Architects' Journal,* 184 (3 Sept. 1986).

35. T. Hardy, *The Return of the Native* (1878; repr. Penguin Classics, 1985).

36. G. Harvey, 'Water Meadow Management the Modern Way', *Farmers' Weekly,* 16 Nov. 1979.

37. M. A. Moon and F. H. W. Green, 'Water Meadows in Southern England', Appendix II to *The Land of Britain, Part 89, Hampshire,* ed.

Dudley Stamp (Geographical Publications, 1940); D. E. Tucker, 'Water meadows', *Water Space,* Winter 1978/9.

38. S. J. Smith, I. Ridge and R. M. Morris, 'The Biomass Potential of Seasonally Flooded Wetlands', in H. Egnéus and A. Ellegard. *Bioenergy 84,* Vol. 2, *Biomass Resources* (Elsevier, 1985). See also M. E. Heath, D. S. Metcalf and R. F. Barnes, *Forages* (Iowa State University Press, 1973).

39. B. J. Whitehead, 'The Topography and History of North Meadow, Cricklade', *Wiltshire Archaeological and Natural History Magazine,* 76 (1982); F. Evans, D. Painter and K. Payne, North Meadow National Nature Reserve: Draft Management Plan (Nature Conservancy Council, unpublished, 1983).

40. In *The Poems of Gerard Manley Hopkins,* ed. W. H. Gardner and N. H. Mackenzie (Oxford University Press, 1981).

41. E. Thomas, 'Haymaking', in *The Collected Poems of Edward Thomas,* ed. R. G. Thomas (Oxford University Press, 1981).

42. R. Mabey, *The Common Ground* (Hutchinson, 1980).

43. Ibid.

44. L. Zhang and H. Hytteborn, 'Effect of Ground Water Regime on Development and Distribution of *Fritillaria meleagris*', *Holarctic Ecology,* 8 (1985).

45. F. Evans, The Snake's-head Fritillary at North Meadow NNR, Wilts (Nature Conservancy Council, Southern Region, internal report, 1983); Evans, Painter and Payne, above, n. 39; F. H. Perring and L. Farrell, *Vascular Plants* (British Red Data Books, SPNC 1, 1977).

46. D. A. Ratcliffe, Objectives and Strategy for Nature Conservation in Great Britain (Nature Conservancy Council, 1983).

47. T. C. E. Wells, A. Frost and S. Bell, *Wildflower Grasslands from Crop-Sown Seed and Hay-Bales* (Nature Conservancy Council, 1986); G. Taylor, ed., Creating Attractive Grasslands: Diverse Swards—the Do's and Don'ts (Bingley National Turfgrass Council, Workshop Report 5, 1985), D. Macintyre, 'The Use of Haybales and Hayfields as Seed Sources' (unpublished paper).

48. *The Times,* 21 Apr. 1986.

49. R. G. Stapledon, 'The Nutritive Influence of the Herbs of Grassland, *Farming,* 2 (1948); idem and D. Wheeler, 'January 16th on the Wiltshire Downs', *Journal of the Ministry of Agriculture,* 58 (1951).

50. B. Thomas and A. Thompson, 'The Ash Content of Some Grasses and Herbs on the

Palace Leas Hayplots at Cockle Park', *Journal of Experimental Agriculture*, 16 (1948).

51. A. Kirchner, 'The Common Dandelion, *Taraxacum officinale*. A Monograph Considered from the Agricultural Standpoint', *Zeitschrift für Acker und Pflanzenbau*, 4 (1955); J. Brüggemann *et al.*, 'The Chemical Composition of Several Species of Permanent Pastures and the Influence of Fertilizers on them', *Bayer. Landw. Jahrb.*, 37 (1960); F. K. Van der Kley, 'On the Variations in Contents and in Inter-Relations of Minerals in Dandelion and Pasture Grass', *Netherlands Journal of Agricultural Science*, 4 (1956).

52. W. Shakespeare, *Cymbeline*, iv. 2.

53. G. M. Hopkins, 'Pied Beauty', see above, n. 40.

54. *The Times*, 26 Aug. 1986.

55. R. Body, *Agriculture: The Triumph and the Shame* (Temple Smith, 1982).

56. *Sunday Times*, 31 Aug. 1986.

57. *The Times*, 28 Jan. 1985.

58. *Hansard*, Standing Committee E, 6 Mar. 1985; *Hansard*, Wildlife and Countryside Amendment Act, 26 Apr. 1985.

59. P. Hughes, 'Foxes in Leicester', *Leicester Wildlife News*, 8 (Summer 1985).

60. A. E. Housman, 'A Shropshire Lad', in *The Collected Poems of A. E. Housman* (Cape, 1939).

61. C. A. Sinker *et al.*, *Ecological Flora of the Shropshire Region* (Shropshire Trust for Nature Conservation, 1985).

62. Maltby, above, n. 28. See also Environmental Data Services Ltd., ENDS Report, 54, Aug. 1980; Somerset County Council, Peat Local Plan, Sept. 1986; P. D. Moore, *European Mires* (Academic Press, 1984).

63. E. A. Woodruffe-Peacock, 'The Ecology of Thorne Waste', *Naturalist*, 1920, 1921; B. C. Eversham, M. Limbert and M. Lynes, 'A Peat Moor Study: Hatfield Moors', *Lapwing, Journal of the Doncaster and District Ornithological Society*, nos. 11–15; B. C. Eversham, 'Hatfield Moors: Classified List of Vascular Plants'; 'Hatfield Moors: Classified List of Hatfield Moors Invertebrates', Jan. 1986, unpublished.

64. See T. A. Rowell, S. M. Walters and H. J. Harvey, 'The Rediscovery of the Fen Violet, *Viola persicifolia* at Wicken Fen, Cambridgeshire', *Watsonia*, 14 (1983); T. A. Rowell, 'The Fen Violet at Wicken Fen', *Nature in Cambridgeshire*, 26 (1983).

65. J. D. Peart and L. A. Rutherford, eds., *A Review of Opencast Coal and its Environmental Impact* (Newcastle-upon-Tyne Polytechnic, Mar.

1986); C. G. Down and J. Stocks, *Environmental Impact of Mining* (Applied Science Publishers, 1977).

66. D. A. H. Brown, 'Compensation for Conservation—A Review of Experience. The Present System—What is its Cost?', in *Proceedings of RASE/ADAS/RICS conference, Feb. 1986;* C. Mathias, Changes in Land Use in England, Scotland and Wales 1985 to 1990 and 2000 (Nature Conservancy Council—Laurence Gould Consultants Ltd., Feb. 1986).

67. *ADA Gazette*, Spring 1985.

## Chapter 6

1. T. S. Eliot, *Collected Poems* (Faber and Faber, 1963).

2. T. Blench, *Regime Behaviour of Canals and Rivers* (Butterworth's Scientific Publications, 1957).

3. F. Severn, 'The Conflict between Land Drainage and Habitat Conservation. The Effect of Watercourse Vegetation on Channel Capacity' (Unpublished thesis, Imperial College of Science and Technology, University of London, 1982).

4. R. D. Hey, 'River Mechanics', *Journal of the Institute of Water Engineers and Scientists*, 40 (1986).

5. K. Taylor, 'The Influence of Watercourse Management on Moorhen Breeding Biology', *British Birds*, 77 (1984).

6. K. Williamson, 'A Bird Census Study of a Dorset Dairy Farm', *Bird Study*, 18 (1971). See also A. E. Smith, 'The Impacts of Lowland River Management', ibid. 22 (1975).

7. S. M. Haslam, 'Changing Rivers and Changing Vegetation in the Past Half-century', proceedings *Aquatic Weeds and their Control* (1981).

8. C. F. Mason, S. M. Macdonald and A. Hussey, 'Structure, Management and Conservation Value of the Riparian Woody Plant Community', *Biological Conservation*, 29 (1984).

9. S. Swales, 'Environmental Effects of River Channel Works Used in Land Drainage Improvement', *Journal of Environmental Management*, 14 (1982).

10. Mason, Macdonald and Hussey, above, n. 8.

11. N. T. H. Holmes, *Wildlife Surveys of Rivers in Relation to River Management* (Water Research Centre, 1986).

12. W. Binder, P. Jürging, *et al.*, 'Natural River Engineering: Characteristics and Limitations', *Garten und Landschaft*, Special Rivers Issue, Feb. 1983; W. Binder, 'Courses of Streams and

Rivers—Their Form and Care', *Garten und Landschaft*, Jan. 1978.

13. G. Lewis and G. Williams, *Rivers and Wildlife Handbook* (Royal Society for the Protection of Birds, 1984). See also C. Newbold, J. Purseglove and N. Holmes, Nature Conservation and River Engineering (Nature Conservancy Council, 1983); *Conservation and Land Drainage Guidelines* (Water Space Amenity Commission, 1980).

14. Institution of Civil Engineers, Power for the Use of Man—Addresses to the 150th Anniversary Celebration of the Royal Charter, in Chilver Report (ICE, 1978).

## Chapter 7

1. M. Drayton, in *The Works of Michael Drayton*, ed. J. William Hebel (Shakespeare Head Press, 1932–41).

2. W. Shakespeare, *Hamlet*, IV. vii.

3. S. Parkinson. *Scenes from the George Eliot Country* (Leeds, 1888).

4. E. Thomas, 'The Brook', in *Collected Poems of Edward Thomas* (Oxford University Press, 1981).

5. D. Glitz, 'Artificial Channels—the "Ox-bow" Lakes of Tomorrow', *Garten und Landschaft*, Feb. 1983.

6. E. Whitehead, A Guide to the Use of Grass in Hydraulic Engineering Practice (Construction Industry Research and Information Association, 1976).

7. J. Realton, 'Breeding Biology in Huntingdonshire Farm Ponds', *British Birds*, 65; S. Peterken, 'Resurvey of Ponds in Huntingdonshire' (Nature Conservancy Council, unpublished report, 1980).

8. M. Wood, 'Artificial Otter Holts: New Objectives for an Old Idea', *Water Space*, 15 (1979).

9. *Literature Survey and Preliminary Evaluation of Streambank Protection Methods* (Chief of Engineers, US Army, Washington, D.C., 1977).

10. *Sunday Times*, 18 Jan. 1970.

11. In *Selected Poems and Prose of John Clare*, ed. E. Robinson (Oxford University Press, 1967).

12. E. S. Edees, *Flora of Staffordshire* (David and Charles, 1972).

13. A. Pope, 'Windsor Forest' (1713), in *The Poems of Alexander Pope* (Methuen, 1963).

14. J. Coles, *The Archaeology of Wetlands* (Edinburgh University Press, 1984).

15. C. French and M. Taylor, 'Desiccation and Destruction: The Immediate Effects of Dewatering at Etton, Cambridgeshire', *Oxford Journal of Archaeology*, 4 (1985).

## Chapter 8

1. R. Weyl, An Ecological Survey of the Wetlands of the Blackwater Catchment (Conservation Branch, Department of Environment, Northern Ireland, 1983).

2. J. Richardson, A. G. Jordan and R. H. Kimber, 'Lobbying, Administrative Reform and Policy Styles. The Case of Land Drainage', *Political Studies*, 26 (1978); A. G. Jordan, J. Richardson and R. H. Kimber, 'The Origins of the Water Act of 1973', *Public Administration*, 55 (1977).

3. D. Baldock, *Wetland Drainage in Europe. The Effects of Agricultural Policy in Four EEC Countries* (International Institute for Environment and Development, 1984); E. C. Penning-Rowsell *et al.*, *Floods and Drainage* (Allen and Unwin, 1986); T. Hollis, 'Land Drainage and Nature Conservation: Is There a Way Ahead?', *ECOS*, 1 (1980).

4. *Hansard*, Written Answers, 11 Dec. 1985.

5. The map of 'Land Drainage in England and Wales' is available from MAFF, Lion House, Willowburn Estate, Alnwick, Northumberland.

6. J. K. Friend and M. J. Laffin, *Organisation and Effectiveness of Internal Drainage Boards* (Tavistock Institute of Human Relations, 1983).

7. C. Pye-Smith and C. Rose, *Crisis and Conservation: Conflict in the British Countryside* (Pelican, 1984); *Hansard*, Nov. 1981.

8. P. Lowe *et al.*, *Countryside Conflicts. The Politics of Farming Forestry and Conservation* (Temple Smith/Gower, 1986). Until the early 1980s, IDBs obtained 50 per cent grant aid from MAFF.

9. See also ch. 3, pp. 69–70.

10. Friend and Laffin, above, n. 6.

11. Pye-Smith and Rose, above, n. 7; *Hansard*, Nov. 1981.

12. Lowe *et al.*, above, n. 8.

13. Friend and Laffin, above, n. 6.

14. Ibid.

15. Pye-Smith and Rose, above, n. 7.

16. *ADA Gazette*, Spring 1983.

17. Pye-Smith and Rose, above, n. 7.

18. C. Addison, *A Policy for British Agriculture* (Gollancz, 1939).

19. R. H. Miers, 'Land Drainage—Its Problems and Solutions' (Paper read at Institute of Water Engineers and Scientists, Summer Conference, 1979).

20. R. H. Best, *Land Use and Living Space* (Methuen, 1981); idem and M. A. Anderson,

'Land Use Structure in Britain 1971–81', *The Planner*, 70 (1984).

21. Food From Our Own Resources (HMSO, Cmnd. 6020, 1975).

22. G. Bennet, 'Bristol Floods 1968. Controlled Survey of Effects on Health of Local Community Disaster', *British Medical Journal*, 3 (1970).

23. N. T. H. Holmes, *Wildlife Surveys of Rivers in Relation to River Management* (Water Research Council, 1986).

24. Baldock, above, n. 3.

25. The widely used definition of wetlands adopted by the Ramsar Convention is: 'areas of marsh, fen, peat land or water, whether natural or artificial, permanent or temporary, with water that is static or flowing, fresh, brackish or salt, including areas of marine water the depth of which at low tide does not exceed six metres'. For our purposes, we should add wet or periodically flooded lowland pasture and areas of reclaimed marsh where the water-table is permanently high.

26. Baldock, above, n. 3.

27. D. A. Scott, *A Preliminary Inventory of Wetlands of International Importance for Wildfowl in West Europe and Northwest Africa* (IWRB, Slimbridge, Glos., 1980).

28. Baldock, above, n. 3.

29. J. Billaud, *Marais Poitevin* (Editions L'Harmattan, 1984).

30. E. Rechard, *Quel avenir pour les Marais Communaux de la partie occidentale du Marais Poitevin?* (Ecole Nationale Supérieure d'Agronomie et des Industries Alimentaires de Nancy, 1980).

31. 'Pour une Poignée de maïs', *Combat Nature, Revue des associations écologiques et de défense de l'environnement*, 66 (1984).

32. Ibid.

33. R. Mabey, *The Common Ground* (Hutchinson, 1980); M. Shoard, *The Theft of the Countryside* (Temple Smith, 1980); E. C. Penning-Rowsell, Proposed Drainage Scheme for Amberley Wildbrooks, Sussex: Benefit Assessment (Middlesex Polytechnic Flood Hazard Research Centre, 1978). This early economics report contains no hint that Amberley was an important SSSI.

34. K. Carlisle, Conserving the Countryside. A Tory View (Conservative Political Centre, 1984).

35. D. Millar, 'Flood Plain Dairy Men Lose Drainage Battle,' *Farmers' Weekly*, Apr. 1978.

36. C. Dickens, *Great Expectations* (1860; repr. Penguin, 1986).

37. D. Jordan, ['A Son of the Marshes'], *The Wild-Fowl and Sea-Fowl of Great Britain* (Chapman and Hall, 1895). See also idem, *Within an Hour of London Town among Wild Birds and their Haunts* (Blackwood, 1892); idem, *Annals of a Fishing Village* (Blackwood, 1892).

38. G. M. Williams, 'The Impact of Land Drainage on the Birds of the North Kent Marshes and a Strategy for Future Management' (unpublished University of London thesis, 1986).

39. E. H. Gilham, J. Harrison, *Wildfowl in Great Britain. Kent,* Nature Conservancy Monograph 3 (HMSO, 1963); E. H. Gilham and R. C. Homes, *The Birds of the North Kent Marshes* (Collins, 1950).

40. K. W. Smith, 'The Status and Distribution of Waders Breeding on Wet Lowland Grasslands in England and Wales', *Bird Study*, 30 (1983).

41. Williams, above, n. 38. See also idem *et al.*, 'The Effects on Birds of Land Drainage Improvements in the North Kent Marshes', *Wildfowl*, 34 (1983).

42. C. Rose, 'A Drain on the Budget', *Sunday Times*, 27 Mar. 1983.

43. S. Connor, 'Digging Holes in Soil Research', *New Scientist*, 104 (1984).

44. K. Jefferies, 'River Idle Washlands SSSI' (Nature Conservancy Council, unpublished report, 1986).

45. Pye-Smith and Rose, above, n. 7.

46. G. Cox and P. Lowe. *A Battle not the War: The Politics of the Wildlife and Countryside Act*, The Countryside Planning Year Book, vol. 4 (Geo-Books, 1983); B. Denyer-Green, *Wildlife and Countryside Act 1981: The Practitioner's Companion* (RICS, 1983).

47. Quoted in P. Lowe *et al.*, above, n. 8.

48. Lowe *et al.*, above, n. 8.

49. W. M. Adams, *Implementing the Act* (BANC and WWF, 1984).

50. *Farming News*, 30 Mar. 1984.

51. W. Cobbett, *Rural Rides* (1830; repr. Penguin, 1967).

52. J. Piper, *Romney Marsh* (King Penguin, 1950). See also R. Ingrams and F. Godwin, *Romney Marsh and the Royal Military Canal* (Wildwood House, 1980).

53. W. Latimer, A Survey of the Dyke Flora of Romney Marsh (Nature Conservancy Council, 1980).

54. F. M. Firth, *The Natural History of Romney Marsh* (Meresborough Books, 1984).

55. J. Sheail and J. O. Mountford, 'Changes in the Perception and Impact of Agricultural Land Improvement: The Post-War Trends in Romney Marsh', *Journal of the RASE*, 145 (1984).

56. P. Evans, 'Report on Sternburg Farm's Land, Walland Marsh' (Kent Trust for Nature Conservation, unpublished report, Feb. 1982).

57. *Observer*, 21 Mar. 1982.

58. A. Nicolson, *Wetland: Life in the Somerset Levels* (Michael Joseph, 1986).

59. M. Williams, *The Draining of the Somerset Levels* (Cambridge University Press, 1970).

60. Nicolson, above, n. 58.

61. E. W. Barnett, Somerset Levels and Moors Plan (Somerset County Council, 1983).

62. Lowe *et al.*, above, n. 8.

63. Environmentally Sensitive Areas in the UK. Citation for Proposed Area No. 6: Somerset Levels (RSPB, Nov. 1986); S. Davies and R. Jarman, Wildlife of the Somerset Levels and Moors (RSPB, RSNC); B. Storer, *Sedgemoor. Its History and Natural History* (David and Charles, 1972).

64. E. W. Barnett, Somerset Levels and Moors Strategy. Framework for Implementation (Somerset County Council, 1984).

65. The Somerset Wetlands Project. A Consultation Paper (Nature Conservancy Council, South-West Region, 1977).

66. R. Baker, 'Lincolnshire Fens in Somerset', *The Somerset Farmer*, 1976.

67. Land Drainage Survey Report (Wessex Water Authority. Local Land Drainage District, 1979).

68. 'Report on Public Enquiry for Proposed North Gloucestershire Internal Drainage District' (unpublished letter to MAFF, 16 May 1972).

69. Nicolson, above, n. 58.

70. Above, n. 63.

71. Memorandum entitled 'Proposed Site of Special Scientific Interest: West Sedgemoor' (NCC, Taunton office, 31 Mar. 1982).

72. B. Lawrence, *Coleridge and Wordsworth in Somerset* (David and Charles, 1970).

73. R. North, *Wild Britain: The Century Book of Marshes, Fens and Broads* (Century, 1983).

74. *Observer* leader, 30 Jan. 1983.

75. The *Guardian*, 23 Feb. 1983.

76. T. Fishlock, 'Letter from Bharatpur', *The Times*, 26 Mar. 1983.

77. Barnett, above, nn. 61, 64.

78. M. Robins, 'Somerset Moors Breeding Birds' (RSPB unpublished report, 1987). M. Robins and R. Green, 'Changes in the Management of Water Levels in the Somerset Levels' (RSPB unpublished report, 1988).

79. O. Sitwell, *Tales my Father Taught me* (Hutchinson, 1962).

80. J. K. Bowers and C. J. Black, The Soar Valley Improvement Scheme: Submission to the House of Lords Select Committee Considering the Severn–Trent Water Authority Bill (CPRE, 1983); J. B. Chatterton, Proof of Evidence: House of Lords Select Committee on the Severn–Trent Water Authority Bill (STWA, 1983); J. K. Bowers, 'Running Soar', *ECOS*, 5, no. 1 (1984); M. Rayner, C. Mathias and C. Green, 'Soar Valley—A Note', *ECOS*, 5, no. 3 (1984).

81. *The Times*, 20 Aug., 21 Sept. 1984.

82. J. Leland, *Itinerary. 1535–1543* (Centaur Press, 1964).

83. C. Caulfield, 'Britain's Heritage of Wildlife Drains Away', *New Scientist*, 3 Sept. 1981; P. Oldfield, 'The Grain Drain', *Guardian*, 10 July 1981; B. Jackman, *The Countryside in Winter* (Century Hutchinson, 1985).

84. Comments on 'North Duffield Carrs Pumping Scheme; The Case for the Drainage Board' (NCC, York, 1984).

85. What Future for Broadland? A Draft Strategy and Management Plan (The Broads Authority, Nov. 1982).

86. Towards a Landscape Strategy for the Broads (Landscape Group Report, Broads Authority, Feb. 1982).

87. *The Times*, 17 Mar. 1984.

88. Report on the Proposed Seven Mile and Berney Levels Drainage Scheme (Report by John Dossor and Partners Company for the Lower Bure, Halvergate Fleet, and Acle Marshes IDB, 19 Apr. 1983).

89. Lowe *et al.*, above, n. 8.

90. Ibid.

91. *The Times*, 18 Feb. 1984.

92. 'Marshes Saved as Whitehall Brawls', *New Scientist*, 18 Nov. 1982.

93. A. Lees, Evidence to the House of Commons Environment Committee on the Wildlife and Countryside Act 1981 (Broadland Friends of the Earth, 1984).

94. E. C. Penning-Rowsell *et al.*, *Floods and Drainage* (Allen and Unwin, 1986).

95. T. O'Riordan, 'Lessons from the Yare Barrier Controversy' (Inaugural Lecture, School of Environmental Sciences, University of East Anglia, Nov. 1980); A. Charnock, 'Flood Scheme would Alter Face of Broads', *New Civil Engineer*, 11 Oct. 1979.

96. This is a convenient abbreviation of its correct title, which is the Lower Bure, Halvergate Fleet, and Acle Marshes IDB.

97. A. Long, The Politics of Wetland Conservation. A Case Study of the Halvergate Marshes, Norfolk (Ecole Polytechnique/CPRE, 1982).

98. 'Decision on Halvergate Marshes' (DoE/MAFF press release, 10 Nov. 1982).

99. Unprecedented Use of Audit Procedures against an IDB (Broadland Friends of the Earth, July 1985).

100. *Hansard*, 4 Apr. 1984.

101. Under article 4 of the General Development Order, to prevent drainage being undertaken.

102. House of Commons Official Report, 12 Nov. 1984.

103. Penning-Rowsell *et al.*, above, n. 94.

104. *Water Bulletin*, 22 Aug. 1986.

105. Financing and administration of Land Drainage, Flood Prevention and Coast Protection in England and Wales (Cmnd. 9449, HMSO, Mar. 1985).

106. Lifting the Veil. CPRE's Response to the Ministry of Agriculture, Fisheries and Food Green Paper on 'Financing and Administration of Land Drainage, Flood Prevention and Coast Protection in England and Wales' (CPRE, Sept. 1985).

107. Ibid., G. Williams, The RSPB's Response to 'Financing and Administration of Land Drainage, Flood Prevention and Coast Protection in England and Wales' (RSPB, Sept. 1985).

108. Friend and Laffin, above, n. 6.

109. Privatisation of the Water Authorities in England and Wales (Cmnd 9734, HMSO, Feb. 1986).

110. The Water Environment: The Next Steps. The Government's Consultative Proposals for Environmental Protection under a Privatised Water Industry (DoE, Welsh Office, 28 Apr. 1986).

111. The National Rivers Authority. The Government's Proposals for a Public Regulatory Body in a Privatized Water Industry (DoE, Welsh Office, 16 July 1987).

112. 'North and South Unite on a Flood Scheme, *New Civil Engineer*, 25 July 1985.

113. 'Black Future for the Blackwater', *Birds*, Winter 1986. G. Williams and D. Browne, 'The Drainage of Northern Ireland', ECOS 8 (3), 1987.

114. Public Expenditure Priorities: Agriculture (Northern Ireland Economic Council, Report 41, Jan. 1984).

115. Above, n. 113.

116. M. Newson, Further Assessment of the Environmental Implications of Land Drainage Improvement in the Catchment of the River Blackwater, Northern Ireland (Report for RSPB from Institute of Hydrology, 1986).

117. Baldock, above, n. 3.

118. P. S. Watson, Habitat and Bird Surveys in the River Blackwater Catchment, Counties Tyrone and Armagh (RSPB, Belfast, 1984); R. Weyl, An Ecological Survey of the Wetlands of the Blackwater Catchment (Department of Environment for Northern Ireland, 1983).

119. W. Blake, 'The Marriage of Heaven and Hell', 1793, in *Blake: Songs of Innocence and of Experience, and Other Works*, R. B. Kennedy, ed. (Macdonald and Evans, 1970).

# INDEX

Figures in **bold** refer to illustrations